Semiconductor Circuits: Worked Examples

Semiconductor Circuits: Worked Examples

BY

J. R. ABRAHAMS, B.Sc. (ENG.), M.Sc., A.M.I.E.E.

AND

G. J. PRIDHAM, B.Sc. (ENG.), A.M.I.E.E., A.M.I.E.R.E.

Senior Lecturers, Enfield College of Technology

PERGAMON PRESS

OXFORD · LONDON · EDINBURGH · NEW YORK
TORONTO · SYDNEY · PARIS · BRAUNSCHWEIG

Pergamon Press Ltd., Headington Hill Hall, Oxford
4 & 5 Fitzroy Square, London W.1
Pergamon Press (Scotland) Ltd., 2 & 3 Teviot Place, Edinburgh 1
Pergamon Press Inc., 44–01 21st Street, Long Island City, New York 11101
Pergamon of Canada, Ltd., 6 Adelaide Street East, Toronto, Ontario
Pergamon Press (Aust.) Pty. Ltd., 20–22 Margaret Street, Sydney, New South Wales
Pergamon Press S.A.R.L., 24 rue des Écoles, Paris 5e
Vieweg & Sohn GmbH, Burgplatz 1, Braunschweig

First edition 1966
Library of Congress Catalog Card No. 66-21145

(2784/66)

Contents

Contributors

THE following members of staff at Enfield College have co-operated in the preparation of this book:

S. H. Bevan, B.SC., DIP.ED. (Lecturer)

K. G. Coe, B.SC. (ENG.), A.M.I.E.E. (Lecturer)

R. B. Matthews, B.SC. (ENG.) (Lecturer)

J. E. Mundy, B.SC., A.M.I.E.E., A.M.I.E.R.E. (Principal Lecturer)

Preface

THIS book has been written as a companion volume to the authors' *Semiconductor Circuits: Theory, Design and Experiment*. Over the past few years many of our students have asked for a book concentrating on transistor questions and we believe this to be the first such to be published in this country.

The questions have been taken from a wide range of examination papers, both at the technician and undergraduate level, and are classified, according to the main topic of the question, into eleven chapters.

Within each chapter those questions having fully worked solutions are grouped in approximate order of increasing difficulty. Where two somewhat similar questions are included (and this applies particularly to the overworked topic of transistor amplifiers), a full solution is given only to the first question, and the second question is put toward the end of the chapter.

For ease of cross-reference the chapter titles are identical with those in *Semiconductor Circuits: Theory, Design and Experiment*. In some cases the reader is referred to the appropriate section in that book. Such a cross-reference to the companion volume would, for example, appear as (*T.D.E.* 3.2) if the reference is to Chapter 3, section 2.

The sources from which the majority of questions were taken are indicated by the following abbreviations, which in every case are followed by the calendar year of the examination.

City and Guilds of London Institute Telecommunications Technician Examinations (third, fourth and fifth years respectively)	C. and G. 3, 4 or 5
Higher National Certificate (Enfield) in Electrical Engineering (first, second and endorsement years, respectively)	H.N.C. 1, 2 or 3
Higher National Diploma (Enfield) in Electrical Engineering (second or third year)	H.N.D. 2 or 3
Institution of Electrical Engineers. Part Three (Graduate) Examination in Electronic Engineering or Applied Electronics	I.E.E.
University of London, B.Sc. in Electrical Engineering (Part 2 or 3)	B.Sc. 2 or 3

A list of questions is given at the start of each chapter and therefore an index has not been included in this book.

Enfield,
August 1964

J.R.A.
G.J.P.

Acknowledgements

WE ARE grateful for the permission of the Senate of London University, the Institution of Electrical Engineers and the City and Guilds of London Institute to use certain questions taken from past examination papers of those three organizations. The full responsibility for the accuracy or otherwise of the worked examples lies with the authors, and the answers are in no way authorized by the examining bodies.

Mr. N. Hiller, who is joint editor of this series, has been very helpful at various stages in the preparation of the manuscript.

We wish to acknowledge the permission of the Governors of Enfield College (through the Head of the Electrical Engineering Department) for this book to be published.

Glossary of Terms

α	Small signal or a.c. current gain (any frequency) in common base.
α_0	Low frequency current gain in common base.
$\bar{\alpha}$	d.c. or large signal current gain in common base.
A	Area (m^2)
β	Small signal or a.c. current gain (any frequency) in common emitter.
β_0	Low frequency current gain in common emitter.
$\bar{\beta}$	d.c. or large signal current gain in common emitter.
B	Magnetic flux density (webers/m^2).
C_e	Capacity across emitter junction (Farads).
C_c	Capacity across collection junction (Farads).
e	Charge on an electron ($1{\cdot}6 \times 10^{-19}$ coulombs).
E	Electric field strength (V/m).
E_H	Hall field strength (V/m).
E_F	Fermi level.
f_α	Cut-off frequency in common base.
f_β	Cut-off frequency in common emitter.
F_E	Force on electron due to electric field (Newtons).
F_H	Force on electron due to magnetic field (Newtons).
G	Conductance (mhos).
$h_{11}, h_{12}, h_{22}, h_{21}$	Hybrid parameters of a transistor in common base.
$h'_{11}, h'_{12}, h'_{22}, h'_{21}$	Hybrid parameters of a transistor in common emitter.

$h''_{11}, h''_{12}, h''_{22}, h''_{21}$	Hybrid parameters of a transistor in common collector.
i_b	Alternating base current.
i_c	Alternating collector current.
i_e	Alternating emitter current.
I_B	Direct base current.
I_C	Direct collector current.
I_{CO}	Leakage current in grounded base.
I'_{CO}	Leakage current in grounded emitter.
I_E	Direct emitter current.
I_L	Direct load current.
I_S	Reverse saturation current.
I_T	Total current.
I_Z	Zener diode current.
J	Current density (A/m^2).
k	Boltzmann's constant ($1{\cdot}37 \times 10^{-23} J/°K$).
K (or S)	Stability factor ($\partial I_c / \partial I_{co}$).
m	Stage gain.
n	Type of semiconductor.
n	Number of "free" electrons per m^3.
n	Number of turns.
N_I	Intrinsic carrier density per m^3.
p	Type of semiconductor.
p	Number of "free" holes per m^3.
ϱ	Resistivity (Ω-m).
r_b	Base resistance (Ω).
r'_{bb}	Extrinsic base resistance (Ω).
r_c	Resistance of reverse biased collector diode (Ω).
r_e	Resistance of forward biased emitter diode (Ω).
R_B	Base resistance (Ω).
R_B	Ballast resistance (Ω).
R_D	Dynamic impedance of a tuned circuit (Ω).
R_H	Hall coefficient (m^3/C).

R_{in}	Input resistance (Ω).
R_L	Load resistance (Ω).
R_{out}	Output resistance (Ω).
R_Z	Slope resistance of a Zener diode (Ω).
S (or) K	Stability factor.
T	Absolute temperature (°K).
μ	Feedback ratio ($\delta V'_{eb}/\delta V'_{eb}$).
μ_e	Electron mobility (m^2/V sec).
μ_h	Hole mobility (m^2/V sec).
v_e	Drift velocity of electrons (m/sec).
v_h	Drift velocity of holes (m/s).
v_{eb}, v_{ce}, v_{cb}	Alternating voltages between emitter, collector and/or base terminals.
V_{EB}, V_{CE}, V_{CB}	Direct voltages between emitter, collector and/or base terminals.
V_{CC}	Transistor supply voltage.
V_F	Forward voltage applied to a *pn* junction.
V_R	Reverse voltage applied to a *pn* junction.
V_Z	Reverse breakdown voltage of a Zener diode.
η	Efficiency.
$y_{11}, y_{12}, y_{22}, y_{21}$	Admittance parameters in common base (mhos).
$y'_{11}, y'_{12}, y'_{22}, y'_{21}$	Admittance parameters in common emitter (mhos).
$y''_{11}, y''_{12}, y''_{22}, y''_{21}$	Admittance parameters in common collector (mhos).
$z_{11}, z_{12}, z_{22}, z_{21}$	Impedance parameters in common base (Ω).
$z'_{11}, z'_{12}, z'_{22}, z'_{21}$	Impedance parameters in common emitter (Ω).
$z''_{11}, z''_{12}, z''_{22}, z''_{21}$	Impedance parameters in common collector (Ω).
Z_{in}	Input impedance (Ω).
Z_L	Load impedance (Ω).
Z_{out}	Output impedance (Ω).

CHAPTER 1

Basic Physical Theory

Q.1.1

Describe with suitable sketches the crystal structure of a pure semiconductor material. Hence explain the meaning of the term covalent bond and discuss its importance in relation to the conduction of electricity in intrinsic, *n*-type and *p*-type semiconductors. Sketch the energy level diagram in each case. Show that the resistivity of an intrinsic semiconductor is given by

$$\varrho = \frac{1}{N_I e(\mu_e + \mu_h)}.$$ [H.N.C. 3, 1964]

A.1.1

The crystal structure of a pure semiconductor material is in the form of a body centred cubic lattice, each atom sharing its outermost valence electrons with four neighbouring atoms. This may be shown in two dimensions by Fig. 1.1.1 with each atom having eight valence electrons under its influence giving the inert gas structure. A three-dimensional sketch is shown in Fig. 1.1.2.

Between each pair of atoms a link is formed. The motion of the shared valence electrons produces a binding force between atoms and under equilibrium conditions this is balanced by electrostatic repulsion between the positively charged nuclei. This link is known as a covalent bond.

For conduction to take place in intrinsic semiconductor material covalent bonds must be broken. At room tempera-

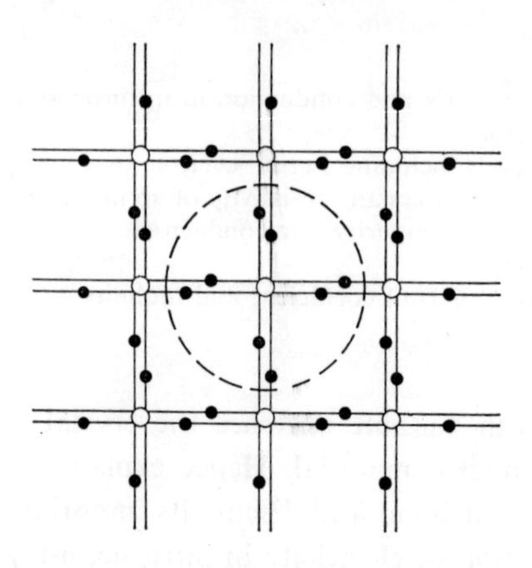

FIG. 1.1.1. Two-dimensional sketch of crystal structure of intrinsic semiconductor.

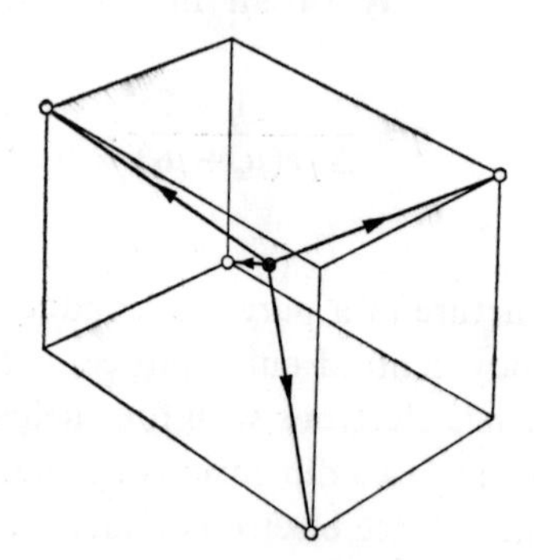

FIG. 1.1.2. Three-dimensional sketch of crystal structure of intrinsic semiconductor.

ture some bands are broken by thermal energy to produce hole-electron pairs. Part of the current flow is then due to electron motion and a smaller fraction due to the motion of holes.

In *n*-type material, where a pentavalent impurity is added to the basic material, four electrons are absorbed into covalent bonds but the fifth gives rise to an electron available for conduction. When a trivalent impurity is added some covalent

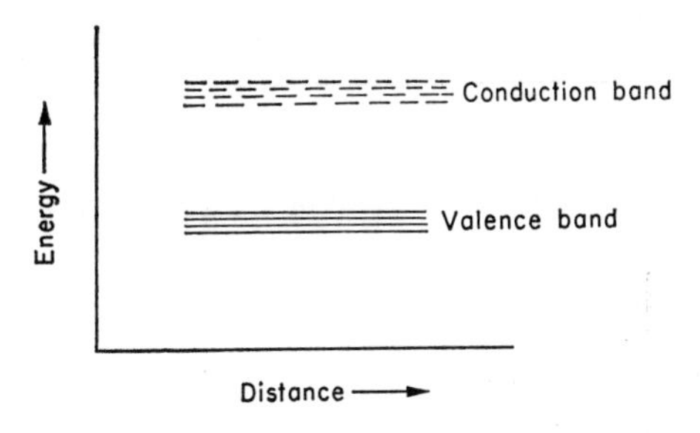

FIG. 1.1.3. Energy level diagram of an intrinsic semiconductor.

bonds are incomplete and conduction in this case is due to the flow of holes. Such material is known as *p* type.

The energy band diagrams for each case are shown in Figs. 1.1.3, 1.1.4 and 1.1.5. Figure 1.1.3 shows the energy level diagram for an intrinsic semiconductor. The valence band is almost full and the conduction band almost empty with an appreciable energy gap between them. This gap is 0·72 eV for germanium and 1·1 eV for silicon. Some electrons are raised from the valence to the conduction band by acquiring thermal energy and these produce electron flow in the conduction band and hole flow in the valence band.

With *n*-type material the electrons outside the covalent bonds give rise to a local energy level (donor level) just below the conduction band. Electrons are easily raised from the donor level into the conduction band to produce the majority (electron) carriers.

If we consider p-type material a local energy level (acceptor level) is produced just above the valence band. Conduction is

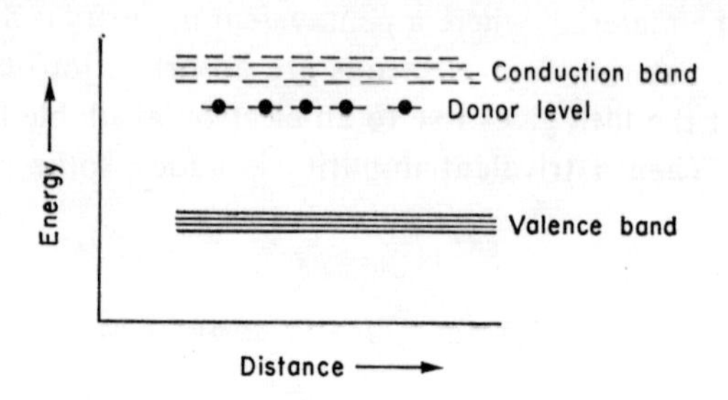

FIG. 1.1.4. Energy level diagram of a n-type semiconductor.

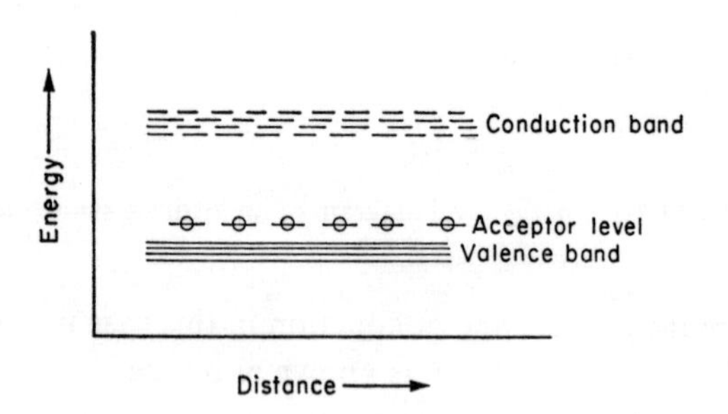

FIG. 1.1.5. Energy level diagram of a p-type semiconductor.

then due to hole flow at the top of the valence band and holes are the majority carriers.

For a specimen of intrinsic material of cross-section A m^2:

Total current = charge passing a given point per second
= charge due to electron flow + charge due to hole flow,

i.e. $$I = N_I e v_e A + N_I e v_h A,$$

where N_I = number of free electrons (or holes) per m^3,
e = charge on an electron (C),

v_e = drift velocity of electrons (m/sec),
v_h = drift velocity of holes (m/sec).

But
$$v_e = \mu_e E,$$
$$= \mu_e \cdot \frac{V}{L},$$

where μ_e = mobility of electron carriers (m^2/V sec),
E = electric field strength (V/m),
V = applied voltage (V),
L = length of specimen (m),

and
$$v_h = \mu_h E,$$
$$= \mu_h \cdot \frac{V}{L},$$

where μ_h = mobility of hole carriers m^2/V sec.

Therefore
$$I = N_I e \mu_e \frac{V}{L} \cdot A + N_I e \mu_h \frac{V}{L} \cdot A,$$
$$\frac{V}{I} = \frac{L}{N_I e(\mu_e + \mu_h) A}.$$

Hence by comparison with the relation $R = \varrho \cdot l/A$.

Resistivity
$$\varrho = \frac{1}{N_I e(\mu_e + \mu_h)} \ \Omega\text{-m}.$$

For further details see *T.D.E.*, Chapter 1.

Q.1.2.

What is meant by the energy level of an electron in an atom? With reference to solids explain the following terms, illustrating your explanation with diagrams where this is possible: filled energy band, conduction band, Fermi level, crystal lattice, valency electrons, covalent bond, intrinsic and impurity semiconductors. [I.E.E., June 1960.]

A.1.2.

Energy Level

The Bohr model of the atom is a useful representation of the atomic structure of materials. The electron orbits are such that centrifugal forces are balanced by electrostatic forces and for simple cases values of potential and kinetic energies may be calculated. Each electron has a definite amount of energy associated with it. This is known as the energy level of the electron and is measured in electron volts.

If we consider a single atom, the electrons are arranged in shells of definite energy level and those nearer the nucleus require more energy to remove them from the parent atom.

Energy Bands

For the complete lattice the electron energies are modified by the fields of adjacent atoms. The single energy levels spread out to form bands of energies. Electrons can move from one level to another within a band fairly easily, by acquiring small

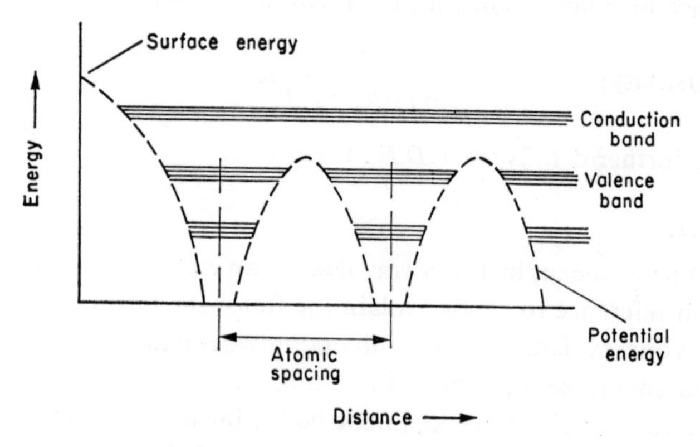

FIG. 1.2.1. Energy band diagram for a solid.

amounts of energy, but appreciable energy is required for the electron to move from one band to another, a typical band structure is shown in Fig. 1.2.1.

The inner energy bands where all the energy levels are completely filled, contain tightly bound electrons. Electrons in these bands have no effect on the chemical and physical properties of the material. A filled energy band is one where there is an electron for every possible energy value within the band.

The outermost band, where electrons are not rigidly bound to the nucleus, is known as the conduction band. Next to this is the valence band. The difference in energy levels between the top of the valence band and the bottom of the conduction band is the fundamental explanation for the difference in behaviour of conductors, insulators and semiconductors.

Fermi Level

If we consider the distribution of electron energies in a crystal, the resulting pattern will be as shown in Fig. 1.2.2. At absolute zero all the lower energy levels, up to a value E_F would

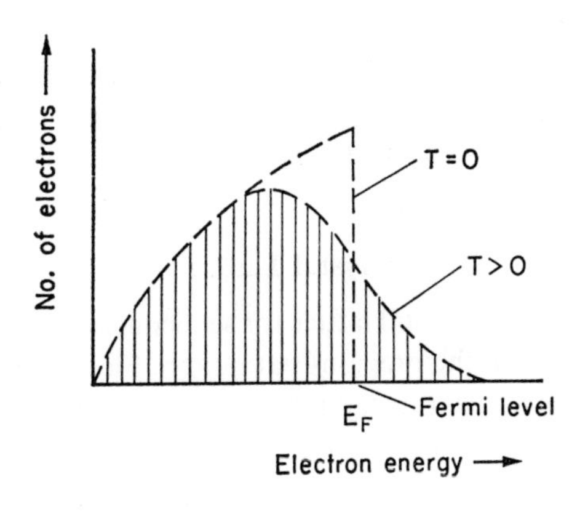

FIG. 1.2.2. Fermi level.

be filled. At any finite temperature some electrons acquire additional thermal energy and jump to a higher energy level, so that there are as many full energy levels above E_F as there are empty energy levels below E_F. The value E_F is known as the Fermi level and may be defined as the energy level at which the probability of finding an electron is one half. The Fermi level is important when considering the action of semiconductor devices.

Valency Electrons

These are the outermost electrons in an atom and take part in chemical and physical reactions. As their name indicates they lie in the valence energy band and must be raised into the conduction band to take part in the process of conduction.

Covalent Bond

In any material some bond exists between neighbouring atoms. The covalent bond is a common type and occurs when electrons are shared between atoms. With semiconductor materials each atom shares its four valence electrons with four neighbouring atoms, so that each atom has under its influence eight electrons giving an inert gas structure. The motion of two electrons about two neighbouring atoms gives rise to a bond between them known as a covalent bond.

Crystal Lattice

The form of the crystal lattice for semiconductors has already been indicated in the previous paragraph. Electrons are shared between atoms forming tight bonds between the atoms. This results in a body-centred cubic lattice, as described in the previous question (A.1.1).

Intrinsic and Impurity Semiconductors

An intrinsic semiconductor is one which can be regarded as a pure material. In such a material the impurity level is extremely low and for conduction covalent bonds must be broken.

An impurity semiconductor is one where a known small amount of pentavalent or trivalent impurity is added to the basic semiconductor material. These impurity atoms are absorbed into the crystal structure, but give rise to either an electron outside the covalent bond (for a pentavalent impurity) or an incomplete bond (for a trivalent impurity). The majority carriers are electrons in the former case and holes in the latter.

Q.1.3

Explain what is meant by the term mobility and how it is related to the total current flowing in a piece of intrinsic semiconductor material.

The resistivity of pure germanium at a particular temperature is 0·52 Ω-m. If it is doped with 10^{20} atoms/m^3 of a trivalent impurity estimate the new value of resistivity. Assume the mobility of holes and electrons are 0·2 and 0·4 m^2/V sec respectively and the charge on an electron is $1{\cdot}6\times10^{-19}$ coulombs.

[H.N.C. 3, 1963]

A.1.3

If a potential difference is applied to a piece of semiconductor material, electrons drift towards the positive electrode with a mean velocity v_e m/sec. At the same time holes drift towards the negative electrode at v_h m/sec. The drift velocity of holes is less than that of electrons since each hole movement corresponds to the breaking of a covalent bond while electrons may move many atomic distances without collision.

The ratio of the drift velocity to the applied field E V/m is a constant for a given material, at a given temperature. This constant μ is known as the mobility of the charge carriers;

i.e. for electrons $$\mu_e = \frac{v_e}{E}$$

and for holes $$\mu_h = \frac{v_h}{E}.$$

The dimensions of μ are $\frac{\text{m}}{\text{sec}} \cdot \frac{\text{m}}{\text{V}} = \text{m}^2/\text{V sec}.$

Hence as shown in A.1.1 for a specimen of intrinsic semiconductor material of length L m and cross-section A m²:

Current $$I = N_I e(\mu_e + \mu_h) AE \text{ amps}$$
$$= \frac{N_I e(\mu_e + \mu_h) AV}{L}$$

where N_I = number of electron (or hole) carriers/m³,
e = charge on an electron C.

Putting $L = 1$, $A = 1$ and rearranging, resistivity

$$\varrho = \frac{1}{N_I e(\mu_e + \mu_h)} \ \Omega\text{-m.}$$

With the values given

$$0{\cdot}52 = \frac{1}{N_I 1{\cdot}6 \times 10^{-19}(0{\cdot}2 + 0{\cdot}4)},$$

i.e. $$N_I = 2 \times 10^{19} \text{ charge carriers/m}^3.$$

If 10^{20} atoms/m³ of a trivalent impurity are added, the number of "free" holes must exceed the number of "free" electrons by this amount,

i.e. $$p - n = 10^{20},$$

where p and n are the number of hole and electron carriers respectively.

Also by the law of mass action†

$$pn = N_I^2$$
$$= 4\times 10^{38}.$$

Therefore

$$p-\frac{4\times 10^{38}}{p} = 10^{20}$$

or $$p^2-10^{20}p-4\times 10^{38} = 0,$$

i.e. $$p = \frac{10^{20}\pm\sqrt{(10^{40}+16\times 10^{38})}}{2}.$$

Taking the positive root

$$p = 1{\cdot}04\times 10^{20}.$$

Therefore $$n = 0{\cdot}04\times 10^{20}.$$

Then total current = electron current + hole current,

i.e. $$I = ne\mu_e AE+pe_h\mu_h AE$$
$$= e(n\mu_e+p\mu_h)A\times\frac{V}{L}.$$

Putting $A = 1$, $L = 1$, gives resistivity

$$\varrho = \frac{V}{I} = \frac{1}{e(n\mu_e+p\mu_h)}.$$

Substituting values:

$$\varrho = \frac{1}{16\times 10^{-19}(0{\cdot}04\times 10^{20}\times 0{\cdot}04+1{\cdot}04\times 10^{20}\times 0{\cdot}2)}$$
$$= 0{\cdot}279\ \Omega\text{-m}.$$

Q.1.4

Assuming that the resistivity of pure silicon is 3000 Ω-m and the mobilities of electrons and holes are 0·12 m²/V sec and 0·025 m²/V sec, respectively, determine the resistivity of a

† For the reader unfamiliar with the Law of Mass Action, further details are available in the companion volume (*T.D.E.* 1.12). An alternative reference is Shockley, W., "Transistor Physics", *Proc. I.E.E.*, Part B, Vol. 103, 1956.

sample of silicon when 10^{19} atoms of phosphorus are added per cubic metre. If 2×10^{19} atoms of boron are now added what is the resultant resistivity? What would be the answers if germanium were the base material? Assume the resistivity of pure germanium is 0·6 Ω-m, electron mobility is 0·36 m^2/V sec and hole mobility is 0·17 m^2/V sec.

A.1.4

As in the previous question for an intrinsic semiconductor,

resistivity $$\varrho = \frac{1}{N_I e(\mu_e+\mu_h)}\ \Omega\text{-m}.$$

Hence for silicon $$N_I = \frac{1}{3000\times1{\cdot}6\times10^{-19}(0{\cdot}12+0{\cdot}025)}$$
$$= 1{\cdot}437\times10^{16} \text{ charge carriers/m}^3.$$

With 10^{19} atoms/m^3 of phosphorus added.

Applying the law of mass action

$$np = (1{\cdot}437\times10^{16})^2$$
$$= 2{\cdot}066\times10^{32},$$

also $$n-p = 10^{19}.$$

Therefore $$n = \frac{2{\cdot}066\times10^{32}}{n} = 10^{19}$$

or $$n^2-10^{19}n-2{\cdot}066\times10^{32} = 0,$$

i.e. $$n = \frac{10^{19}\pm\sqrt{(10^{38}+8{\cdot}264\times10^{32})}}{2}.$$

Taking the positive root

$$n = 10^{19} \quad \text{very nearly},$$
p is negligible.

Therefore $$\varrho = \frac{1}{ne\mu_e}$$
$$= \frac{1}{10^{19}\times1{\cdot}6\times10^{-19}\times0{\cdot}12}$$
$$= 5{\cdot}21\ \Omega\text{-m}.$$

If 2×10^{19} atoms of boron are now added per m^3,

$$p-n = N_A - N_D$$
$$= 10^{19},$$

where N_A and N_D are the number of acceptor and donor atoms added per m^3 and

$$pn = 2{\cdot}066 \times 10^{32}.$$

Hence, solving as before, $p = 10^{19}$, n is negligible.

Therefore $$\varrho = \frac{1}{10^{19} \times 1{\cdot}6 \times 10^{-19} \times 0{\cdot}025}$$
$$= 25\ \Omega\text{-m}.$$

The corresponding values for germanium may be determined by the reader. In this case both holes and electrons must be considered and the solutions are $0{\cdot}55\ \Omega$-m and $0{\cdot}64\ \Omega$-m.

Q.1.5

Give and account of the Hall effect and derive an expression relating the Hall voltage in metals to the applied current and magnetic field.

What information about the properties of semiconductors can be obtained by measurements of the Hall voltage under differing physical conditions?

[B.Sc. 2, 1960]

A.1.5

If a current of I amps flows in a conductor, of length L m and cross-sectional area A m^2, lying in a magnetic field of B webers/m^2, the free electrons will experience a force. This is shown in Fig. 1.5.1 and the value of the force is given by

$$F_H = Bev_e \text{ Newtons,}$$

where e = charge on an electron in C,
v_e = drift velocity of electrons m/sec.

The electrons tend to congregate at the base of the conductor giving rise to an electric field E_H V/m across the conductor. The production of this field is known as the Hall effect.

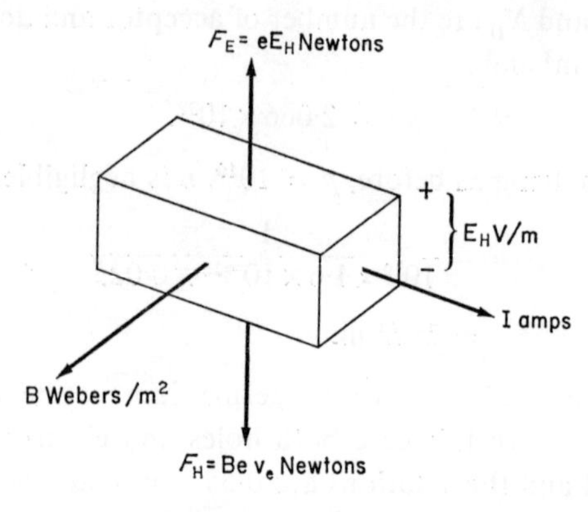

FIG. 1.5.1. Hall effect.

Equilibrium conditions are attained when the force due to the electric field F_E balances that due to the magnetic field,

i.e. $$eE_H = Bev_e.$$

Also the current density $$J = \frac{I}{A} \quad \text{A/m}^2$$

$$= nev_e,$$

where n is the number of free electrons per m³.

Dividing $$\frac{eE_H}{J} = \frac{B}{n}$$

or $$\frac{1}{ne} = \frac{E_H}{BJ} = R_H \quad \text{m}^3\text{/C}.$$

R_H is known as the Hall coefficient and may be determined experimentally.

Since the resistivity of the material is given by

$$\varrho = \frac{1}{ne\mu_e}\ \Omega\text{-m},$$

therefore electron mobility

$$\mu_e = \frac{1}{ne\varrho} = \frac{R_H}{\varrho}\quad \text{m}^2/\text{V sec.}$$

Similar conditions apply to semiconductors, but in this case the Hall e.m.f. is much higher and to calculate the Hall coefficient a constant must be introduced. This is due to the different distribution of velocities in semiconductors and for an *n*-type semiconductor the Hall coefficient is given by:

$$R_H = \frac{3\pi}{8}\cdot\frac{1}{ne}\quad \text{m}^3/\text{C.}\dagger$$

For *p*-type materials a negative sign is introduced and the number of "free" holes used.

Measurement of the Hall voltage, and hence the Hall e.m.f. enable certain physical properties to be determined, i.e.

(a) drift velocity,
(b) carrier concentration,
(c) mobility.

Except for high values of applied electric field strength or magnetic field strength the drift velocity is proportional to the electric field strength but the carrier concentration and mobility are constant.

Measurement of Hall coefficient and resistivity over a range of temperature enables the effect of temperature on carrier concentration and mobility to be determined. The carrier

† The significance of this expression is covered in the companion volume (*T.D.E.* 1. 12).

concentration will increase with temperature due to the thermally produced carriers while the mobility will be reduced due to increased atomic vibration.

Q.1.6

Show that in the case of a conductor the Hall coefficient is given by

$$R_H = \frac{1}{ne} = \frac{E_H}{BJ}.$$

How is this expression modified in the case of a semiconductor?

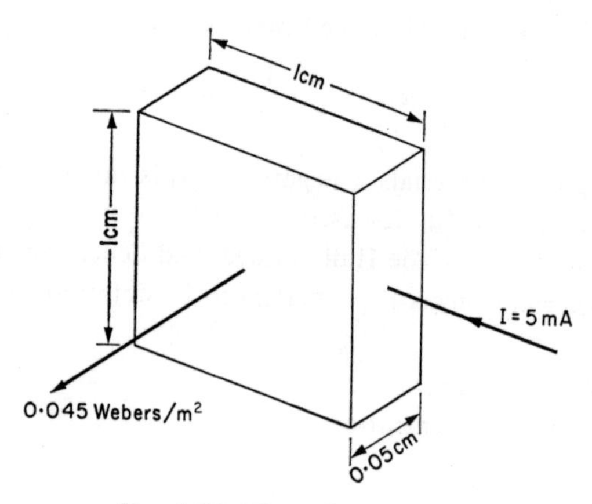

FIG. 1.6.1. Figure for Q.1.6.

Figure 1.6.1 shows a sample of *n*-type germanium. The current of 5 mA flows from a 1·35 V supply and the Hall voltage developed across the specimen is 20 mV. Determine the Hall coefficient and electron mobility for germanium.

[H.N.C. 3, 1964]

A.1.6

The first part of this question is answered in A.1.5.

For the specimen shown:

$$\text{Resistivity} \quad \varrho = \frac{RA}{L}$$

$$= \frac{1\cdot35}{5\times10^{-3}}\times\frac{1\times0\cdot05\times10^{-4}}{10^{-2}}$$

$$= 0\cdot135\ \Omega\text{-m.}$$

$$\frac{1}{ne} = \frac{E_H}{BJ}$$

$$= \frac{20\times10^{-3}/10^{-2}}{0\cdot045\times5\times10^{-3}/0\cdot05\times10^{-4}}$$

$$= 0\cdot044\ \text{m}^3/\text{C.}$$

Therefore Hall coefficient

$$R_H = \frac{3\pi}{8}\times0\cdot044$$

$$= 0\cdot0524\ \text{m}^3/\text{C.}$$

Electron mobility

$$\mu_e = \frac{R_H}{\varrho} = \frac{0\cdot0524}{0\cdot135}$$

$$= 0\cdot39\ \text{m}^2/\text{V sec.}$$

CHAPTER 2

Physical Principles of Semiconductor Devices

Q.2.1

Explain the diffierences between intrinsic, *p*-type and *n*-type conduction in a semiconductor such as germanium.

Describe the mechanism of rectification at *pn* junction. What factors determine the reverse current in such a rectifier?

[B.Sc. 3, 1958]

A.2.1

In intrinsic germanium the carriers available for conduction are due to the breaking of covalent bonds. Both holes and electrons contribute to the total current flow. For *p* type the majority of the current is due to positively charged hole carriers, and is associated with electron movement at the top of the valence band. With *n*-type material the majority of the current is carried by electrons and is associated with the movement of electrons at the bottom of the conduction band.

This subject is covered in more detail in A.1.1.

The mechanism of rectification at a *pn* junction may be explained using the energy level diagrams discussed in A.1.1.

When a *pn* junction is formed electrons flow from the donor levels in the *n*-type material to unoccupied acceptor levels in the *p*-type material, giving rise to a region of intrinsic semiconductor with very few charge carriers. This is the depletion layer.

Electrons flow until the raising of the energy level diagram of the negatively charged *p*-type region coupled with a fall in the energy level diagram of the *n*-type region produces an energy barrier sufficient to stop further diffusion of electrons.

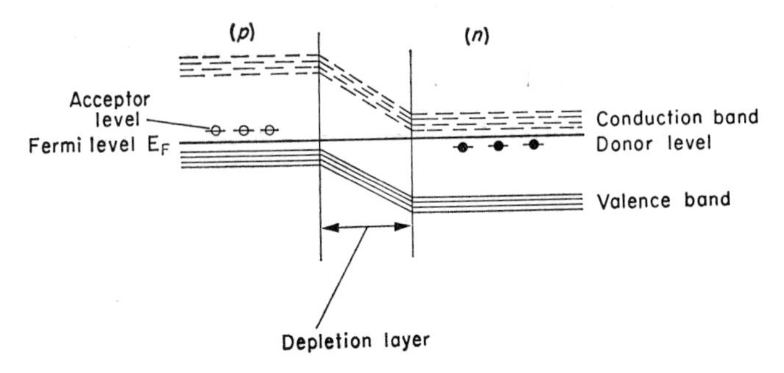

FIG. 2.1.1. Energy level diagram for a *pn* junction with zero bias.

This corresponds to the Fermi levels E_F being continuous from one side of the junction to the other. This is the condition for zero bias shown in Fig. 2.1.1.

With forward bias, i.e. the *p* side made positive with respect to the *n*, the energy barrier is lowered by an amount eV_F joules where V_F is the forward bias applied. This results in the flow of majority carriers trying to maintain the barrier. Electrons flow from the *n*-type and holes from the *p*-type giving rise to an appreciable current. The energy level diagram for this condition is shown in Fig. 2.1.2.

Reverse bias results in an increase in the energy barrier and the only current that flows is due to thermally produced minority carriers (i.e. holes in the *n*-type and electrons in the *p*-type material). A typical characteristic is shown in Fig. 2.1.3. Such a characteristic shows a rapid increase of current with applied voltage in the forward direction and the reverse current quickly reaching its saturation value I_S. This characteristic would lead to rectification of an alternating input.

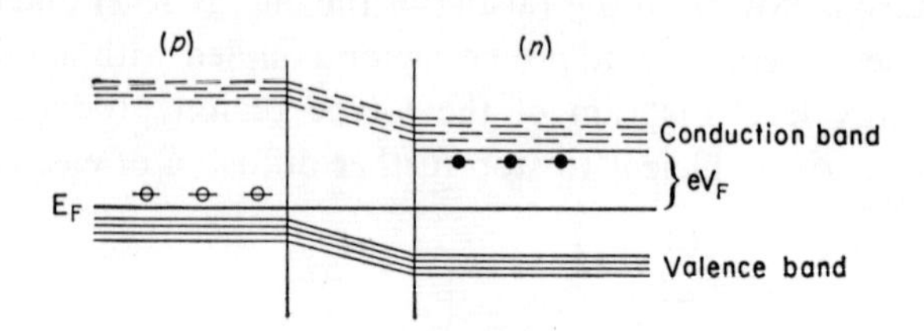

FIG. 2.1.2. Energy level diagram for a *pn* junction with forward bias.

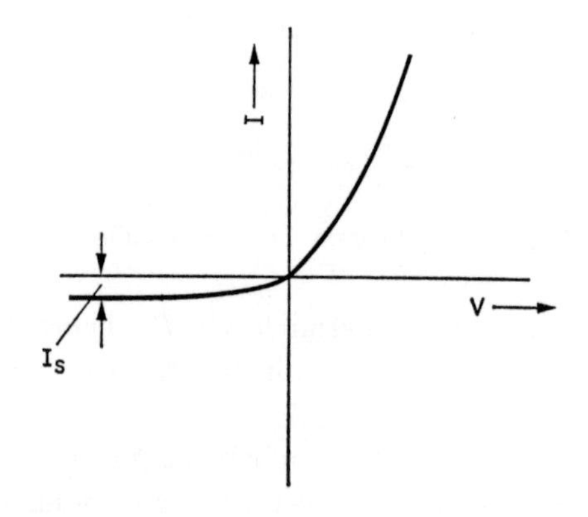

FIG. 2.1.3. Characteristic of a *pn* diode.

As indicated above, the reverse current I_S, being due to thermally produced minority carriers, is dependent on the ambient temperature and the power dissipated in the device. The ability to produced hole–electron pairs thermally is also a major consideration and in this respect silicon is superior to germanium due to the larger energy gap between valence and conduction bands. Above a certain reverse voltage Zener breakdown, avalanche breakdown, or a combination of the two will take place and the reverse current flow will be limited by the resistance of the external circuit.

Q.2.2

Discuss the effects of the amount of impurity added on the characteristics of *pn* diodes. Both the forward and reverse characteristics should be considered.

A.2.2

The usual form of *pn* diode characteristic is as shown in A.2.1 (Fig. 2.1.3). The characteristic shows a rapid increase in forward current with an increase in forward voltage but a very small reverse current. The impurity content of such diodes is about one part in 10^7 or 10^8. The characteristic is not valid for high reverse voltages when breakdown occurs or large forward currents where high charge concentrations are involved.

With a higher impurity content about 1 part in 10^5 the forward characteristic is of the same form but the reverse characteristic exhibits a definite breakdown at low voltage. Such diodes are known as Zener diodes with Zener breakdown occurring at up to about -6 V. A typical characteristic is shown in Fig. 2.2.1. Zener diodes should not be confused with voltage reference diodes that break down at high reverse voltages. The latter depend on the avalanche effect.

When the impurity content is about 1 part in 10^3 or 10^4 the depletion layer formed at the junction is very narrow. For small values of forward bias, full energy levels are separated from empty energy levels by the width of the depletion layer and electrons can "tunnel" through the energy barrier. Such diodes are known as "tunnel" or "Esaki" diodes. The resultant current is a combination of "tunnel" and diffusion currents giving a characteristic as shown in Fig. 2.2.2. The important part of this characteristic is the negative resistance portion which finds many applications in modern circuitry. When a reverse voltage is applied Zener breakdown occurs at a very low reverse voltage.

For further details see *T.D.E.*, Chapter 2.

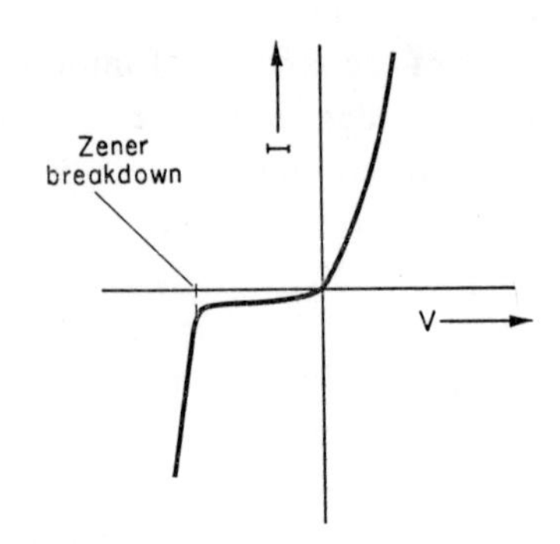

FIG. 2.2.1. Characteristics of a zener diode.

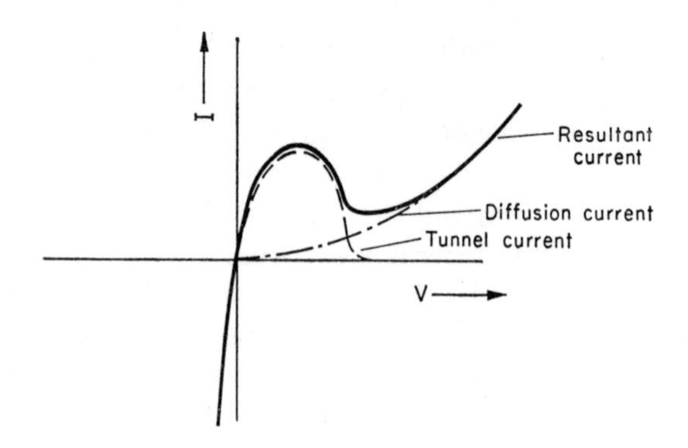

FIG. 2.2.2. Characteristic of a "tunnel" diode.

Q.2.3

Explain what is meant by the energy gap of a semiconductor. Describe with the aid of energy level diagrams the rectification process which occurs at a junction between a *p*-type and an *n*-type semiconductor. Indicate the nature of the expression which relates the current flowing across the junction to the potential difference between the materials, and explain the material properties on which the expression depends.

[B.Sc. 3, 1959]

A.2.3

The energy gap of a semiconductor may be defined as the range of energies between the top of the valence band and the bottom of the conduction band. The only electrons having energies within this gap are those produced by impurity atoms.

This is covered in more detail in A.1.1.

The mechanism of rectification is covered in A.2.1.

The expression for the current I flowing across a semiconductor junction when a voltage V is applied is given by

$$I = I_S(\varepsilon^{eV/kT} - 1),\dagger$$

where e is the charge on an electron in C, k is Boltzmann's constant joules/°K, T is the absolute temperature °K, I_S is the reverse saturation current.

The only factor on the right-hand side of this equation that depends on the material properties is the reverse saturation current I_S.

This leakage current is produced thermally and hence depends on the energy gap of the material. Silicon with an energy gap of 1·1 eV has a much lower leakage current than germanium with its energy gap of 0·72 eV.

Defects in the lattice structure could affect the leakage current but its effect should be negligible in a practical diode.

Q.2.4

Explain using the energy band theory of solids the operation of a *pnp* transistor connected in the grounded base configuration.

† This expression is developed in many contemporary textbooks dealing with the physics of semiconductors.

A.2.4

The energy levels for *p*- and *n*-type semiconductors have been discussed in A.1.1 and the mechanism of rectification in A.2.1. This question extends the use of energy band diagrams to the explanation of transistor operation.

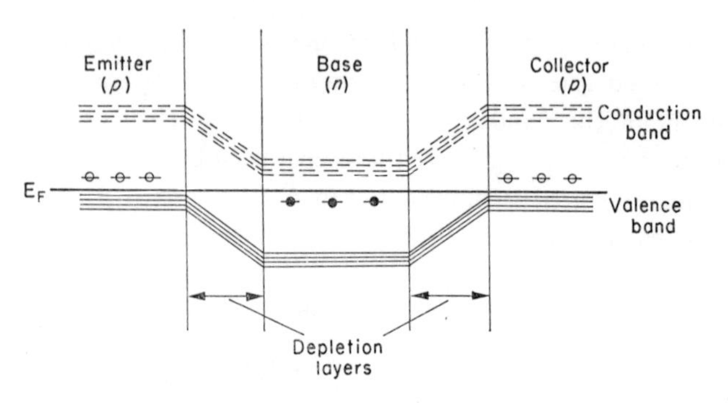

FIG. 2.4.1. Energy level diagram for a *pnp* transistor with zero bias.

A transistor may be regarded as two diodes back to back and the energy level diagram for a transistor with zero bias is shown in Fig. 2.4.1. Under these conditions majority carriers have diffused across the junctions, formed the depletion layers and built up energy barriers between emitter and base and collector and base. Except for thermally produced minority carriers, i.e. electrons in the *p*-type and holes in the *n*-type regions, no current flows.

When normal bias is applied, i.e. the emitter diode forward biased and collector junction reverse biased the energy level diagram is as shown in Fig. 2.4.2. The emitter energy level diagram is lowered by an amount eV_{EB} joules while the collector diagram is raised by an amount eV_{CB} joules, where V_{EB} is the positive potential difference between the emitter and base and V_{CB} the negative potential difference between collector and base.

Under these conditions holes flow across the emitter junction and diffuse across the base before appearing as minority carriers at the collection junction. Some holes recombine with electrons in the base region but the majority are swept across the collector depletion layer to form the collector current. To

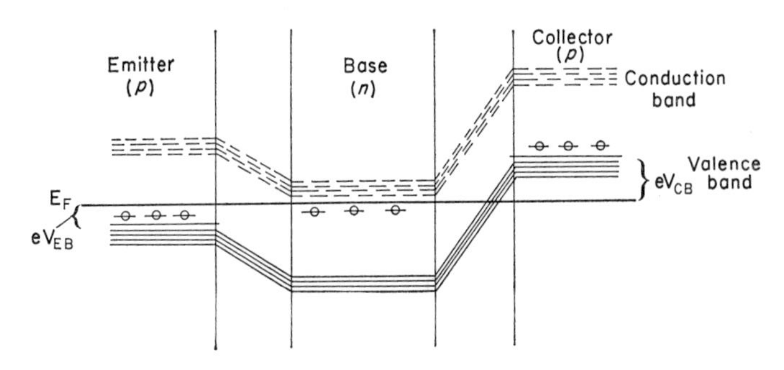

FIG. 2.4.2. Energy level diagram for a *pnp* transistor with normal bias.

reduce the recombination in the *n*-type base region it is made as narrow as possible and the impurity content is much less than in the *p*-type emitter and collector regions. The recombination in the base region is balanced by a small electron current I_B flowing into the base. The low impurity content of the base also means that the number of electrons crossing the emitter junction is very much less than the hole current and the emitter efficiency (i.e. the ratio of hole current to total current) is very high.

In addition to the large fraction $\bar{\alpha}$ of the emitter current I_E that appears at the collector, a small leakage current I_{CO} flows across the collector junction. Hence the d.c. currents are given by the equations:

$$I_C = \bar{\alpha} I_E + I_{CO},$$

$$I_E = I_B + I_C.$$

The d.c. currents and voltages are shown in Fig. 2.4.3.

Small variations in the input current require very little power, due to the low input impedance, but due to the high impedance of the output circuit considerable power can be transferred to a high impedance load. The a.c. power gain of the device is then approximately $\alpha^2 R_L/R_{in}$ where the load resistance R_L may be fifty times the input resistance R_{in} and α about 0·98. α is the small signal current gain given by $\alpha = \delta I_C/\delta I_E$.

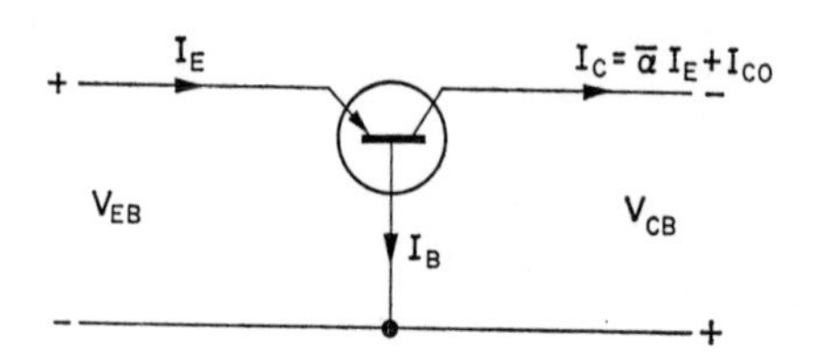

FIG. 2.4.3. Currents and voltages for a *pnp* transistor.

The conversion of d.c. input power into a.c. output power by the use of very little a.c. input power is a fundamental property of the device and leads to its use in many amplifier and oscillator circuits.

Q.2.5

Using as many as possible of the terms listed below, write an essay on semiconductors, junction diodes, and junction transistors. Intrinsic and impurity semiconductors.

Base, emitter, collector, donor, acceptor, *p*-type, *n*-type, majority and minority carriers, lifetime, depletion layer, short-circuit current gain.

[I.E.E., Nov. 1961]

A.2.5

A wide variety of answers is possible for this question and a possible solution is given below.

Semiconductor materials may be roughly defined as materials whose resistivity lies somewhere between that of conductors and insulators. The two elements used commercially in the construction of diodes and transistors are germanium and silicon. These tetravalent elements are characterized by a diamond-type crystal structure, with the valence electrons being shared between atoms. The covalent bonds between atoms absorb all the electrons and there are few carriers available for conduction. For conduction to take place in such intrinsic semiconductors covalent bonds must be broken.

At room temperature thermal energy is responsible for the breaking of some bonds and a small current flows. The breaking of a bond produces a hole–electron pair and in addition to the electron or negatively charged carrier, a hole or positively charged carrier is produced.

The conductivity of semiconductors may be increased by adding pentavalent or trivalent impurities. If a small amount of pentavalent or donor impurity is added the resulting impurity semiconductor has sufficient electrons to fill the covalent bonds and in addition electrons only loosely bound to their parent impurity atoms. These electrons are the majority current carriers in such an *n*-type semiconductor.

In *p*-type semiconductors a trivalent or acceptor impurity is added. This results in incomplete covalent bonds in the semiconductor and the movement of electrons from bond to bond gives rise to a hole moving through the semiconductor. This corresponds to a positive charge movement.

For ideal *p*- and *n*-type semiconductors all the current would be carried by hole and electron carriers respectively. These are the majority carriers in the two types. In practice hole–electron pairs are continually being produced thermally and some "free" electrons will exist in the *p*-type and some "free" holes in *n*-type semiconductors. These are minority carriers and although their lifetime is limited due to recombination they will contribute to the current flow in impurity semiconductors.

Junctions between *p*- and *n*-type semiconductor materials possess the property of rectification and are the basis of the many devices commercially available. At the junction diffusion takes place, "free" electrons moving into the *p*-type and "free" holes into the *n*-type regions. Recombination takes place and a region is formed where there are very few charge carriers. This is the depletion layer and under stable conditions the energy barrier developed across it is sufficient to stop further migration of majority carriers. Forward bias reduces the energy barrier allowing the flow of majority carriers, but reverse bias increases the barrier and the only current flow is due to thermally produced minority carriers.

Transistors have two such junctions, the first forward biased and the second reverse biased. This splits the device into three regions, the first where the current carriers are introduced being known as the emitter, the middle region the base and the output side the collector. The majority carriers from the emitter cross the emitter junction and diffuse across the base region appearing as minority carriers at the collector junction. Recombination is small due to the relatively low impurity content of the base region. The current carriers are swept across the collector depletion layer and a current flows from the collector that is very nearly the same as that introduced at the emitter.

Typically, the short circuit current gain is 0·98 and since the load impedance is usually much less than the output impedance practical current gains are of the same order. Due to the low input and high output impedances appreciable power gain is possible.

Note. In this question an answer has been given that is an attempt at a solution compatible with the 30–40 min available in an examination.

CHAPTER 3

Construction and Characteristics of Transistors

Q.3.1

Sketch labelled diagrams showing a cross-section of the semiconductor material and the connecting leads for the following types of transistor:

alloy junction,

grown junction,

alloy-diffused mesa,

planar epitaxial.

Give a typical dimension for the base width in each case. State briefly the properties required for successful operation with reasonable gain at high frequencies. Compare the four transistors on this basis.

[H.N.C. 3, 1963]

A.3.1

Cross-section diagrams of the semiconductor material and connecting leads for four types of transistor are given in Fig. 3.1.1, in the following order:

(a) alloy junction *pnp*,
(b) grown junction *npn*,
(c) alloy-diffused mesa *pnp*,
(d) planar epitaxial *npn*.

The properties required for reasonable gain at high frequencies may be summarized as follows:

(1) Narrow base region to give minimum transit time of charge carriers.
(2) Low electrical noise figure, to allow for amplification of small signals.
(3) Small area collector–base junction, manufactured to closely defined limits, to give a minimum junction capacitance.
(4) Minimum self inductance and capacitance of leads and encapsulation.

Of the four transistors sketched in Fig. 3.1.1 (c) and (d) both fulfil the first and third of the above requirements. The fourth requirement does not depend, to any great extent on the way in which the semiconductor material is manufactured. However, the planar epitaxial type of transistor has several advantages when compared with the diffused mesa transistor. The junctions are sealed under an inert layer, to give very small leakage currents, and low noise figures. This follows from the fact that the junctions are not exposed to contamination during manufacture and are sealed when the transistor is being

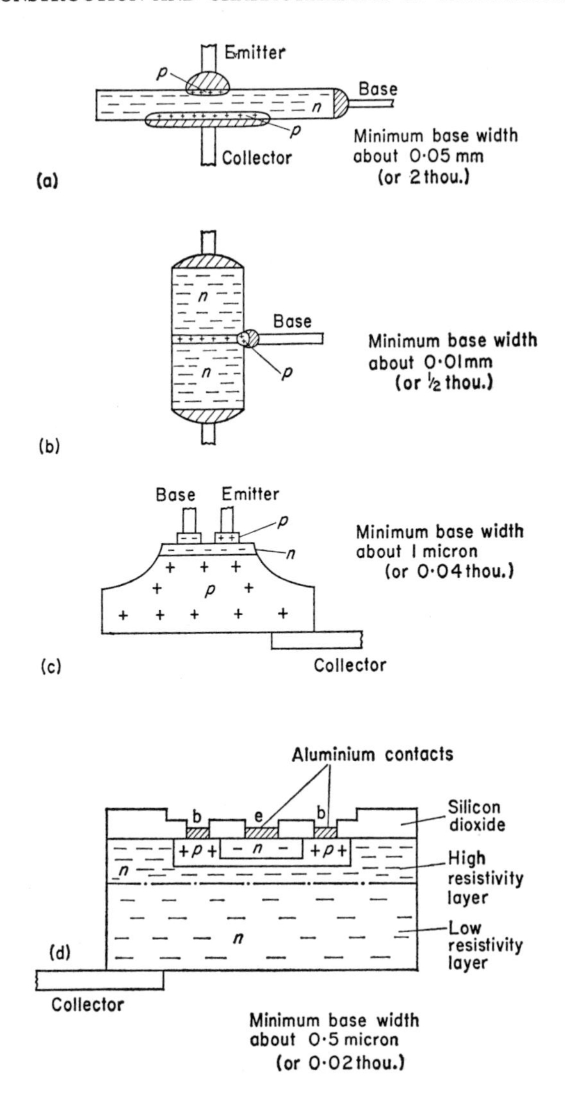

FIG. 3.1.1. Cross-sections of four types of transistor: (a) alloy junction, *pnp;* (b) grown junction, *npn;* (c) alloy-diffused Mesa, *pnp;* (d) planar epitaxial, *npn.*

encapsulated. Further, the bottoming voltage is low and high gain may be achieved with collector currents of less than 0·1 mA.

Q.3.2

Describe methods used to obtain (a) the output, (b) the input characteristics of a low power *pnp* transister operating in the common emitter connection. Include in the description any special requirements in the choice of apparatus and the method of connection. What changes should be made to conduct a similar test on an *npn* transistor? Give typical characteristic curves for the *pnp* connection.

What effect have the input characteristics on the way in which transistors are used?

[H.N.D. 3]

A.3.2

All measurements made on transistors involve either small currents or voltages or the combination of both. While meters are readily available to record such values, their effect on the circuit is often considerable. Testing methods must therefore

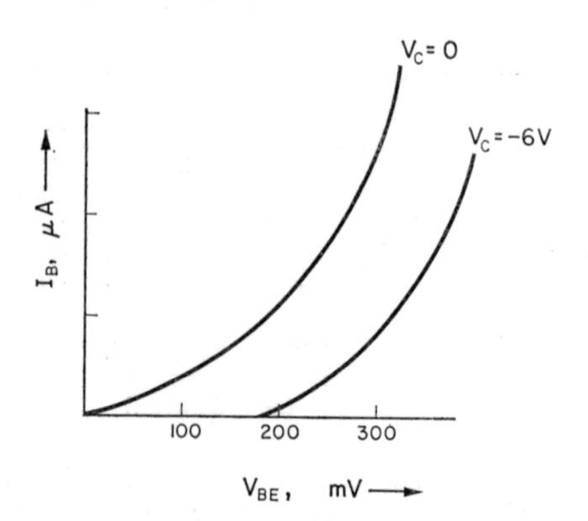

FIG. 3.2.1. Input characteristics of low-power *pnp* transistor.

take this into account, both in the choice of apparatus and their positions in the circuit.

Typical characteristic curves for a low power *pnp* transistor are given in Figs. 3.2.1 and 3.2.2. These may be obtained from the circuit shown in Fig. 3.2.3.

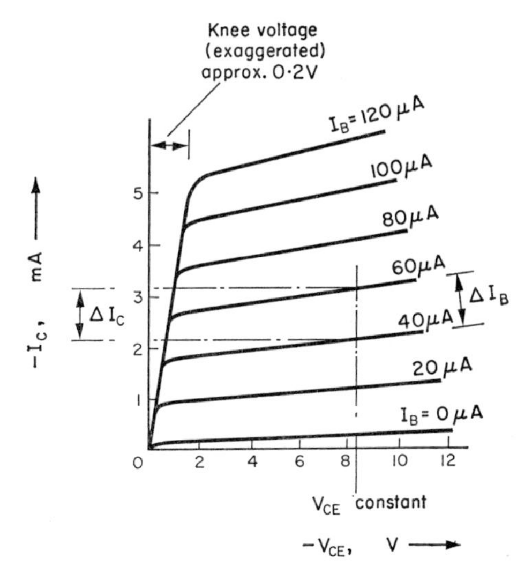

FIG. 3.2.2. Output characteristics of low-power *pnp* transistor.

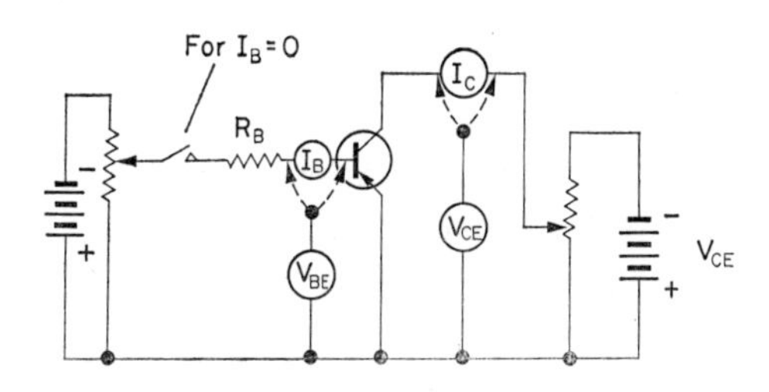

FIG. 3.2.3. Circuit to measure d.c. characteristics of transistors.

The following points should be noted:

(1) D.C. valve voltmeters reading fractions of a volt are not usually available and a 100,000 Ω/V meter will draw 10 μA when reading 1 V. This may be larger than the transistor current.

(2) Sensitive multi-range meters and micro-ammeters have a high series resistance when measuring low currents, causing voltage drops of the order of a volt.

The above problems become serious when trying to find the "knee" of the output characteristic. The positions of meters reading the base and collector voltages must be chosen to minimize these errors and suitable corrections should be made.

(3) Increments of applied voltage should not be equal and extra readings will be required for low collector voltages.

(4) The transistor may heat up during the test and readings should be taken quickly.

(5) Changes of meter range should be recorded and graphs drawn during the course of the experiment.

The circuit itself is straightforward, facilities for varying base and collector voltages being provided. Additional variable resistors may be required for "fine" control but R_B should not be reduced to zero due to the danger of accidental overload.

Tests on an *npn* transistor would be carried out with the polarities of the supplies reversed and the meters reversed. It is important to note that the transistor may be ruined if the polarity of the voltage applied is incorrect.

The input characteristics show a non-linear relationship between I_B and V_{BE}. Thus an applied voltage of sinusoidal waveform will not produce a base and collector current of the same waveform. The input to the base must therefore be from a high impedance or current source, since the transistor acts as a current amplifier.

(N.B. Valves are essentially voltage amplifiers.)

Q.3.3

Distinguish between the ON, OFF and ACTIVE states of a junction transistor in common emitter connection.

Sketch the $I_B - V_{BE}$ characteristic for negative V_{CE} and comment on the magnitude of the leakage current obtained with V_{BE} positive.

[H.N.C. 2]

A.3.3

The output characteristics for a transistor in the grounded emitter configuration are shown in Fig. 3.3.1.

Region I represents the OFF region. In practice a small collector current (a few microamps in a small germanium transistor) does flow. In this case both the junctions are reverse biased.

Region II is the ACTIVE region. In this region the base current controls the collector current.

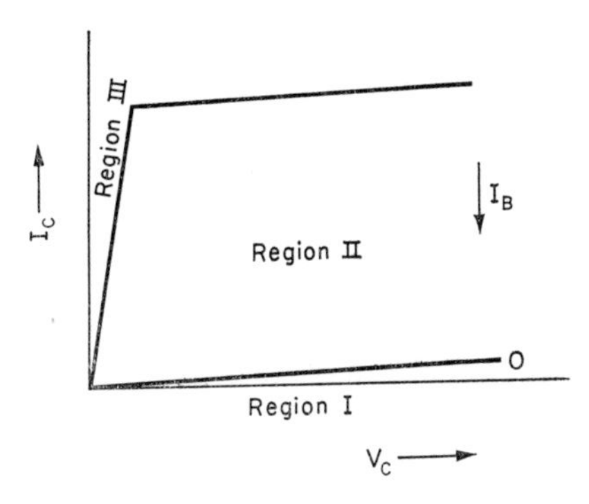

FIG. 3.3.1. Idealized transistor characteristics, to show OFF, ACTIVE and ON regions.

Region III corresponds to the ON region. In practice the transistor would be bottomed. This is the condition when the transistor is conducting heavily and the current is limited by the external circuit. Under these circumstances base, collector and emitter are virtually at the same potential.

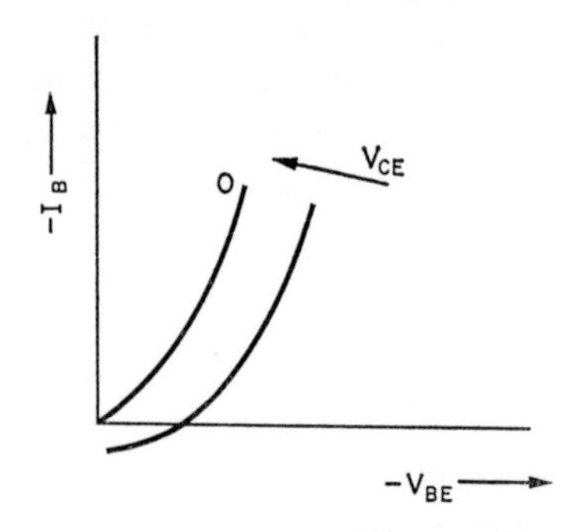

FIG. 3.3.2. I_B–V_{BE} characteristics for A.3.3.

The $I_B - V_{BE}$ characteristics for negative values of V_{CE} are shown in Fig. 3.3.2. There is a slight shift as V_{CE} is made more negative, due to the variation in the width of the collector depletion layer.

The leakage current that flows through a reverse biased junction depends on the thermally produced minority carriers. Its value for the common emitter connection may be much greater than for the common base connection. This is discussed in detail in A.6.5.

Q.3.4

Discuss the factors which influence the current amplification factor of a *pnp* junction transistor. Derive from first principles the relation between the current amplification factors for the grounded base and grounded emitter configurations.

Measurements made on a junction transistor at 30 kc/s gave values of 0·98 and 40 for the magnitudes of the current amplification factors in the grounded base and grounded emitter configurations respectively. Calculate the phase angle of the grounded emitter current amplification factor at 30 kc/s and estimate the cut-off frequency for both configurations.

[B.Sc. 3, 1960]

A.3.4

The current amplification factor of a *pnp* junction transistor depends primarily on the degree of recombination of charge carriers (holes and electrons) which takes place in the base region. The factors which influence this recombination and the current gain are, briefly, as follows:

(a) *Thickness of the base region.* The effective base layer should be much thinner than the mean free path for the majority charge carriers, which in the case of the *pnp* transistor are holes from the emitter.

(b) *The area and nature of the surface of the base region.* A good deal of recombination occurs with impurity substances at the surface of the base, and to minimize this effect the base surface area should be small and passivated. A good example of this is the planar epitaxial transistor (see A.3.1).

(c) *The impurity content of the transistor regions.* The mean free path in the base is dependent on the nature and impurity content of the base material. The current gain is also improved by having a high ratio of majority to minority charge carriers (i.e. a low resistivity emitter and high resistivity base region).

(d) *Frequency of operation.* The potential drops in a correctly biased transistor occur very close to the junctions, at either side of the base region, and the movement of charge carriers across the base is mainly due to drift. The resulting drift velocity is relatively low, giving a significant transit time. High frequency operation therefore results in a lower current gain, for a given base geometry.

(e) *Collector efficiency.* In order that practically all of the majority charge carriers leaving the emitter arrive at the collector, the collector–base junction should be a large proportion of the total base surface area, and much larger than the emitter–base junction area. This is clear in three of the transistors illustrated in Fig. 3.3.1.

The simple definitions for low frequency current gain in common base and common emitter configurations are:

$$\alpha_0 = \left|\frac{\delta I_C}{\delta I_E}\right|_{V_{CB}\ \text{constant}} \quad \text{and} \quad \beta_0 = \left|\frac{\delta I_C}{\delta I_B}\right|_{V_{CE}\ \text{constant}}$$

Hence

$$\frac{\alpha_0}{\beta_0} = \frac{\delta I_B}{\delta I_E}.$$

Also

$$I_B = I_E - I_C,$$

and, therefore,

$$\delta I_B = \delta I_E - \delta I_C,$$

giving that

$$\frac{\alpha_0}{\beta_0} = 1 - \frac{\delta I_C}{\delta I_E} = 1 - \alpha_0$$

or

$$\beta_0 = \frac{\alpha_0}{1-\alpha_0}.$$

Now

$$|\alpha| = \left|\frac{\alpha_0}{1+jx}\right| = 0{\cdot}98$$

and

$$|\beta| = \left|\frac{\alpha_0}{1+jx-\alpha_0}\right| = 40$$

at 30 kc/s,

where

$$x = \frac{f}{f_\alpha},$$

f = the frequency being considered,
f_α = the cut-off frequency in common base.

Hence it is valid, to about $0{\cdot}5\,f_\alpha$ to say that

$$\frac{\alpha_0}{\sqrt{(1+x^2)}} = 0{\cdot}98$$

and
$$\frac{\alpha_0}{\sqrt{[(1-\alpha_0)^2+x^2]}} = 40.$$

Inverting and squaring, we then have that

$$\left(\frac{1}{0{\cdot}98}\right)^2 = \frac{1+x^2}{\alpha_0^2}$$

and
$$\left(\frac{1}{40}\right)^2 = \frac{(1-\alpha_0)^2+x^2}{\alpha_0^2}$$

or
$$\frac{1}{1600} = 1-\frac{2}{\alpha_0}+\left(\frac{1+x^2}{\alpha_0^2}\right) = 1-\frac{2}{\alpha_0}+\left(\frac{1}{0{\cdot}98}\right)^2.$$

Therefore
$$\frac{2}{\alpha_0} = 1+\left(\frac{1}{0{\cdot}98}\right)^2-0{\cdot}000625,$$

where
$$\left(\frac{1}{0{\cdot}98}\right)^2 = 1{\cdot}0404,$$

giving that
$$\frac{2}{\alpha_0} = 2{\cdot}0404-0{\cdot}0063 = 2{\cdot}034.$$

Hence
$$\frac{1}{\alpha_0} = 1{\cdot}017 \quad \text{and} \quad \alpha_0 = \frac{1}{1{\cdot}017}.$$

Using the binomial theorem,

$$\alpha_0 = 1-0{\cdot}17+0{\cdot}0003 = 0{\cdot}983.$$

Therefore
$$\sqrt{(1+x^2)} = \frac{\alpha_0}{\alpha} = \frac{0{\cdot}983}{0{\cdot}98} = 1{\cdot}0029,$$

giving that
$$x = 0{\cdot}076$$

and
$$f_\alpha = \frac{30}{x} = 395 \text{ kc/s (say, 400 kc/s)}.$$

Now f_β (cut-off frequency with common emitter) $= f_\alpha(1-\alpha_0)$. That is $f_\beta = 400\,(1-0{\cdot}983) = 6{\cdot}8$ kc/s.

Finally at 30 kc/s the phase angle associated with the common-emitter current gain is given by

$$\beta \angle \phi = \frac{\alpha_0}{(1-\alpha_0)+jx}, \quad \text{hence} \quad \tan \phi = \frac{-x}{1-\alpha_0}.$$

That is, $$-\phi = \tan^{-1}\frac{0{\cdot}076}{0{\cdot}017} = \tan^{-1} 4{\cdot}47 = 77°.$$

The phase angle of the grounded emitter current gain at 30 kc/s is thus 77 degrees lagging.

Q.3.5

A *pnp* transistor, with a common emitter, short circuit, current gain of 40, has a leakage current (grounded base, open-circuited emitter) of 30 μA. It is used in the circuit of Fig. 3.5.1. Calculate the collector current.

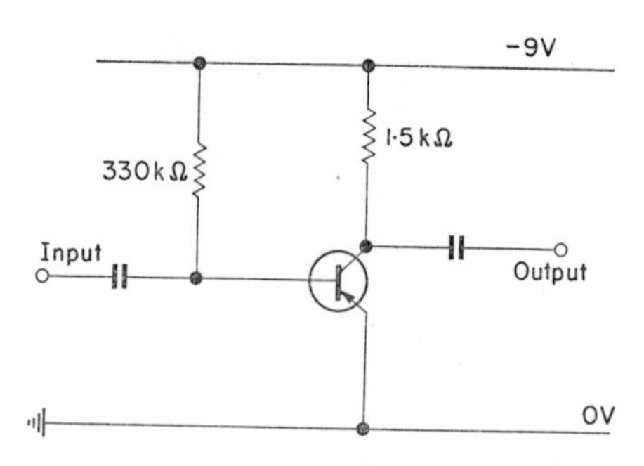

FIG. 3.5.1. Amplifier circuit for Q.3.5.

A.3.5

The value of the collector load resistance is not used in the calculations. It is included in order that it may be assumed that I_C is independent of V_{CE} over the range considered.

Thus the base current (assuming V_{BE} negligible)

is $$I_B = \frac{9 \times 10^6}{330 \times 10^3} = 27\ \mu A$$

and $$I_C = \beta I_B + I'_{CO},$$

where $$I'_{CO} = \frac{I_{CO}}{1-\alpha} = \frac{\beta I_{CO}}{\alpha}.$$

Now $$\beta = \frac{\alpha}{1-\alpha}, \quad \text{giving} \quad \alpha = \frac{40}{41}.$$

Substituting for β, I_B, I_{CO} and α,

we have $$I_C = 40 \times 27 + 40\left(\frac{41}{40}\right) 30\ \mu A$$

or $$I_C = (1{\cdot}08 + 1{\cdot}23)\ \text{mA} = 2{\cdot}31\ \text{mA}.$$

Note

(a) The leakage current exceeds the "signal" current.

(b) The accuracy of the answer depends more upon the components than on the number of significant figures used in the calculations. The resistor values are probably within 10%, while β and I_{CO} are subject to wide variations.

(c) When a collector current of 2·31 mA flows through the collector load of 1·5 kΩ, the collector voltage becomes $V_{CE} = 9-(2{\cdot}31 \times 1{\cdot}5) = 5{\cdot}5$ V.

Thus the original assumption regarding the V_{CE}/I_C relationship is justified.

Q.3.6

The theoretical current/voltage characteristic of a semiconductor junction diode may be represented by the expression

$$I = I_0[\varepsilon^{eV/kT} - 1],$$

where I_0 = junction, saturation current in amps,
k = Boltzmann's constant $1{\cdot}37 \times 10^{-23}$ joule/°K,
V = applied voltage,
T = temperature in degrees K,
e = electronic charge $1{\cdot}60 \times 10^{-19}$ coulombs.

For $I_0 = 0{\cdot}1$ A, calculate the maximum forward current if the temperature is 27°C and the permissible dissipation is 250 mW.

Sketch the current/voltage characteristic, and on the same diagram sketch the practical characteristic, mentioning the reasons for any differences.

[I.E.E., Nov. 1963]

A.3.6

This question is best solved graphically. The point of intersection of the static characteristic and maximum power line giving the maximum current.

Tables to construct these curves are shown in Tables 3.1 and 3.2.

$$\varepsilon^{eV/kT} = \varepsilon^{(1\cdot6\times10^{-19}V)/(1\cdot37\times10^{-23}\times300)} = \varepsilon^{39V}$$

TABLE 3.1

V(mV)	39 V	ε^{39V}	$I = 0\cdot1(\varepsilon^{39V} - 1)$
70	2·73	15·33	1·433
80	3·12	22·64	2·164
90	3·51	33·47	3·247
100	3·9	49·40	4·840

For a maximum dissipation curve for 250 mW.

TABLE 3.2

V (mV)	I (amp)
70	3·58
80	3·13
90	2·78
100	2·50

Hence the graphs of Fig. 3.6.1 can be drawn and from them it is seen that the maximum forward current is 2·85 A.

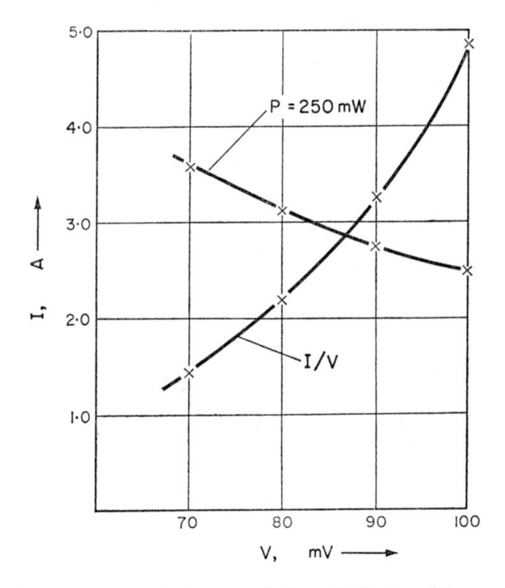

FIG. 3.6.1. Calculated characteristics of diode with a maximum dissipation line for 250 mW.

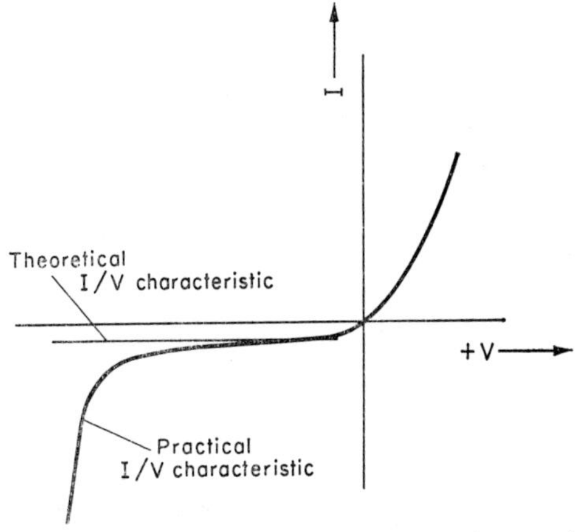

FIG. 3.6.2. Practical and theoretical diode characteristics.

The practical and theoretical characteristics of a semiconductor diode are shown in Fig. 3.6.2.

The rapid increase in reverse current when a certain breakdown voltage is reached is not predicted by the theory of the *pn* junction and it may be due to one of three effects:

(a) Zener breakdown

This is due to a large reverse voltage producing an electric field in the depletion layer that is sufficiently large to liberate electrons from the covalent bonds. These act as charge carriers and produce a rapid increase in the current.

(b) Avalanche breakdown

This is an ionization effect and occurs when the electrons moving through the depletion layer have sufficient energy to ionize atoms by collision. The process is cumulative and will result in a large increase in current.

(c) Thermal breakdown

This can occur when the applied voltage is less than that required for either Zener or avalanche breakdown. As the temperature of the junction is increased additional hole–electron pairs are generated. This action is also cumulative and will result in a large increase in current if the junction temperature becomes sufficiently high.

The slight difference between ideal and practical characteristics in the forward direction, due to the physical resistance of the semiconductor material, is usually negligible.

Q.3.7

Explain why the current amplification factor of a grounded base junction transistor varies with frequency. What types of transistor construction are used to give improved performance at high frequencies?

The current amplification factor α of a grounded base junction transistor, operating at frequency f, is given approximately by the expression

$$\alpha = \frac{\alpha_0}{1+j(f/f\alpha)},$$

where α_0 is the low-frequency amplification factor, and f_α is the alpha cut-off frequency.

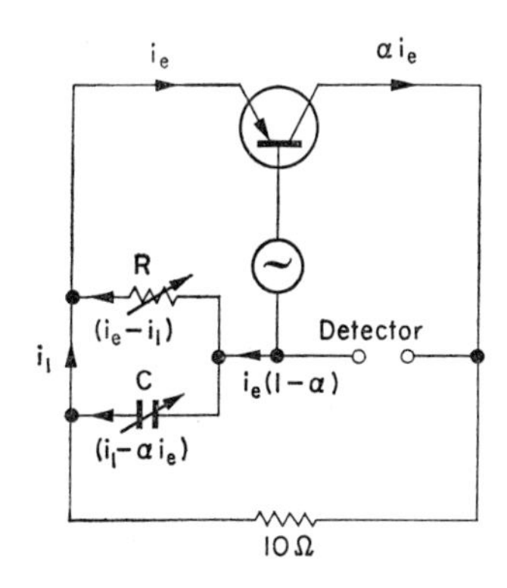

FIG. 3.7.1. Circuit to measure current gain at various frequencies.

Derive from first principles a corresponding expression for the current amplification factor when the transistor is used in the grounded emitter configuration and express the corresponding cut-off frequency in terms of f_α and α_0.

The circuit shown in Fig. 3.7.1 can be used to measure α as a function of frequency. Show that if R and C are adjusted to give zero output in the detector, then

$$\alpha = \frac{1}{1+(10/R)+10j\omega C}$$

provided that the usual assumptions are made concerning the relative magnitudes of the transistor parameters.

[B.Sc. 3, 1960]

A.3.7

Two factors which affect the current gain of a transistor are the transit time of minority carriers through the base and the capacitances at the emitter and collector junctions.

Since the diffusion process is a comparatively slow one it results in a fall in amplification of high frequency signals. The capacitances at emitter and collector junctions shunt the input and output circuits and therefore reduce the amplification as frequency is increased.

Transit time can be reduced by decreasing the width of the base but this cannot be made too thin because of the "punch-through effect". A reduction in transit time is achieved in the drift transistor by manufacturing it with an impurity gradient in the base region which has the effect of accelerating the electrons as they pass through the base. The process also results in an increase in the width of the collector base depletion layer which reduces the collector capacitance.

Another high frequency transistor is the surface barrier transistor. In this an extremely thin base region is achieved by an etching process and the emitter and collector are then formed by electroplating.

It is shown in A.3.4 that the current amplification factor of a grounded emitter transistor is

$$\beta = \frac{\alpha}{1-\alpha}.$$

Therefore $$\beta = \frac{\alpha_0/[1+j(f/f_\alpha)]}{1-[\alpha_0/1+j(f/f_\alpha)]} = \frac{\alpha_0}{(1-\alpha_0)+j(f/f_\alpha)}.$$

The value of β will have fallen 3 dB at some frequency f_β such that

$$1-\alpha_0 = \frac{f_\beta}{f_\alpha},$$

i.e. $$f_\beta = (1-\alpha_0)f_\alpha .$$

Therefore $$\beta = \frac{\alpha_0/(1-\alpha_0)}{1+j[f/(1-\alpha_0)f_\alpha]} = \frac{\beta_0}{1+j(f/f_\beta)}.$$

If in the circuit shown in Fig. 3.7.1 no current flows in the detector, we then have

$$(i_e - i_1)R = \alpha i_e 10, \tag{1}$$

$$(i_1 - \alpha i_e)\frac{1}{j\omega C} = \alpha i_e\, 10. \tag{2}$$

From (1) $$i_e = \frac{Ri_1}{R-10\alpha},$$

(2) $$i_e = \frac{i_1}{\alpha + j\omega C\times 10\alpha}.$$

Therefore $$\frac{Ri_1}{R-10\alpha} = \frac{i_1}{\alpha+j\omega C\times 10\alpha},$$

$$R\alpha + j\omega CR\times 10\alpha = R-10\alpha,$$

$$\alpha[1+(10/R)+j\omega C\times 10] = 1,$$

$$\alpha = \frac{1}{1+(10/R)+j\omega C\times 10}.$$

CHAPTER 4

Equivalent Circuits

Q.4.1

Describe the mechanism of operation of a transistor. From consideration of the physical processes involved derive an equivalent circuit which will represent a transistor. Explain why the current amplification factor varies with frequency.

[B.Sc. 3, 1963]

A.4.1

The mechanism of operation of a transistor has been discussed in A.2.4 and will not be considered here, except where it affects the development of an equivalent circuit.

A transistor comprises low resistivity emitter and collector regions separated by a relatively high resistivity base. The physical resistance of the first two regions may be neglected, but that of the base must be allowed for and the first step in the development of the circuit is shown in Fig. 4.1.1. A resistance $r_{bb'}$, known as the extrinsic base resistance, is connected to an ideal transistor.

The forward biased emitter junction and reverse biased collector junctions are shown by resistances r_e and r_c, but current and voltage generators are required to complete the equivalent circuit. The collector current flows from the high impedance reverse biased collector diode and this is represented by

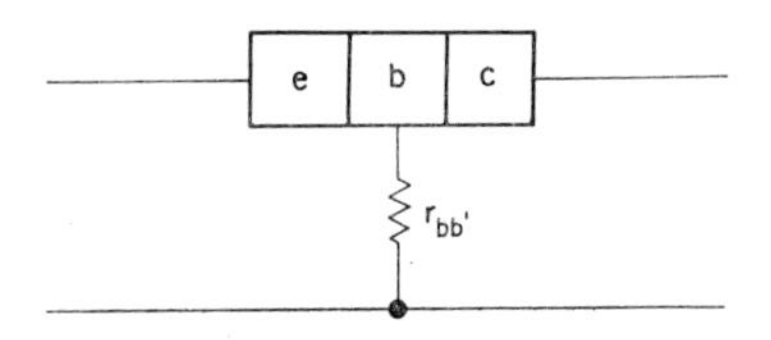

FIG. 4.1.1. Ideal transistor with extrinsic base resistance.

a current generator $\alpha_0 i_e$ across the collector resistance r_c. In practical circuits the collector voltage varies due to the voltage drop across any load that is connected. This varies the thickness of the depletion layer at the collector junction and alters the effective base width. This in turn modifies the amount of

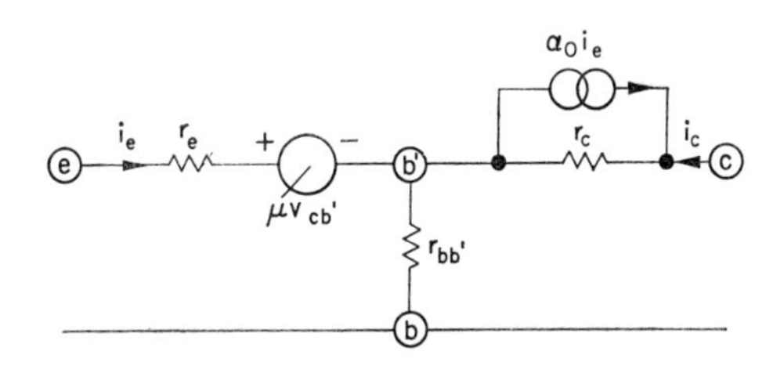

FIG. 4.1.2. An equivalent *T* circuit.

recombination in the base region and is shown by a voltage generator $\mu v_{cb'}$ in the emitter lead. Hence the complete low frequency equivalent circuit is shown in Fig. 4.1.2.

If we now consider the input circuit

$$
\begin{aligned}
v_{eb} &= i_e r_e + \mu v_{cb'} + (i_e + i_c) r_{bb'} \\
&= i_e r_e + \mu (i_c + \alpha_0 i_e) r_c + (i_e + i_c) r_{bb'} \\
&= i_e r_e + (i_c + i_e)(\mu r_c + r_{bb'})
\end{aligned}
$$

since $\alpha_0 \triangleq 1$.

Hence, Fig. 4.1.2 may be transformed to Fig. 4.1.3. The total resistance in the shunt arm $\mu r_c + r_{bb'}$ is now denoted by r_b to give the usual low frequency equivalent circuit.

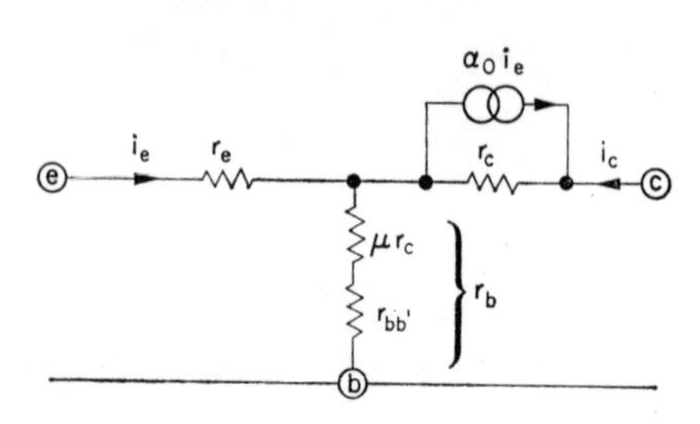

FIG. 4.1.3. Normal equivalent T circuit.

For high frequencies the circuit capacitances must be included, i.e. the collector depletion layer capacitance and the emitter capacitance. The latter includes the depletion layer capacitance and the much larger diffusion capacitance and the equivalent circuit is as shown in Fig. 4.1.4.

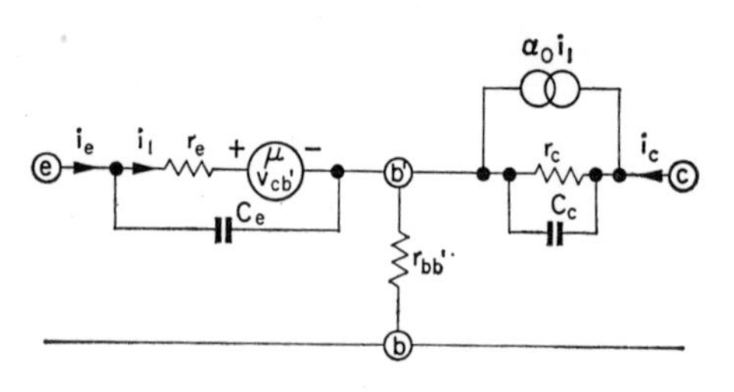

FIG. 4.1.4. h.f. equivalent T circuit.

The current generator is now $\alpha_0 i_1$ where i_1 is the current flowing through the emitter resistance itself.

Then if $\mu v_{cb'}$ is small

$$i_1 r_e = (i_e - i_1)\frac{1}{j\omega C_e},$$

i.e. $$i_1 = \frac{i_e}{1+j\omega C_e r_e}$$

and $$\alpha_0 i_1 = \frac{\alpha_0 i_e}{1+j\omega C_e r_e} = \alpha i_e$$

where $$\alpha = \frac{\alpha_0}{1+j\omega C_e r_e}.$$

This shows that the current amplification factor varies with frequency. The variation is due mainly to the diffusion capacitance and is therefore dependent on the transit time of charge carriers across the base of the transistor. For further details see *T.D.E.*, sections 4.1 and 4.2.

Q.4.2

Draw an equivalent circuit, valid at low frequencies for a junction transistor used as a small signal amplifier, in the common emitter connection. Explain briefly the significance of the parameters used. One or two major factors is likely to limit the h.f. response in this connection, depending on the magnitude of the load resistance. State what these factors are, and give a simplified analysis of the effect of either one of them justifying any assumption made.

[I.E.E., June 1958]

A.4.2

The low frequency equivalent *T* circuit of a transistor has been derived in A.4.1 and Fig. 4.1.3. and may be reorientated to give the common emitter equivalent circuit shown in Fig. 4.2.1.

At high frequencies circuit capacitances have to be considered and the major factors likely to limit the h.f. response are:

(1) The collector depletion layer capacitance in parallel with load circuit.

(2) The reduction in current gain due to the transit time effect.

Considering the latter case, it has been shown in A.4.1 that

$$\alpha = \frac{\alpha_0}{1+j\omega C_e r_e},$$

where α_0 is the current gain at l.f., α is the current gain at a frequency $f = \omega/2\pi$, C_e is the emitter capacitance, r_e is the emitter resistance, i.e.

$$\alpha = \frac{\alpha_0}{1+j(\omega/\omega_\alpha)} \quad \text{where} \quad \omega_\alpha = \frac{1}{C_e r_e}$$

$$= \frac{\alpha_0}{1+j(f/f_\alpha)}.$$

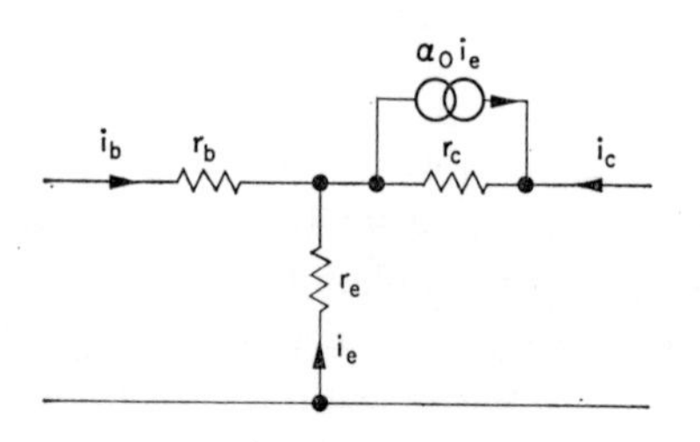

Fig. 4.2.1. Grounded emitter equivalent *T* circuit.

f_α is known as the α cut-off frequency and may be defined as that frequency where the current gain falls to $1/\sqrt{2}$ of the l.f. value.

If we assume the load resistance is small compared with the output resistance, for a common emitter circuit

$$\text{current gain} = \frac{i_c}{i_b}$$

$$= \frac{i_c}{i_e - i_c}$$

$$= \frac{\alpha}{1-\alpha}$$

$$= \beta.$$

Then
$$\beta = \frac{\alpha_0/1+j(f/f_\alpha)}{1-[\alpha_0/1+j(f/f_\alpha)]}$$
$$= \frac{\alpha_0}{1-\alpha_0+j(f/f_\alpha)}$$
$$= \frac{\alpha_0/1-\alpha_0}{1+j(f/f_\alpha)\times 1/(1-\alpha_0)}$$
$$= \frac{\beta_0}{1+j(f/f_\alpha)(1+\beta_0)},$$
since
$$\beta_0 = \frac{\alpha_0}{1-\alpha_0}.$$

This equation shows that there is a very rapid fall off in current gain in the common emitter configuration.

Q.4.3

Explain why h parameters are the most commonly measured parameters of a junction transistor.

Draw a circuit diagram of apparatus for measuring the h parameters at a frequency of 1000 c/s. Explain how the measurements are carried out and how the results are calculated.

What other measurements would be required in order to evaluate the high-frequency performance of the transistor?

[B.Sc. 3, 1959]

A.4.3

The hybrid parameters of a transistor are shown in Fig. 4.3.1 and the circuit equations are
$$v_1 = i_1 h_{11} + h_{12} v_2$$
and
$$i_2 = v_2 h_{22} + h_{21} i_1$$
where v_1, i_1, v_2 and i_2 are small changes in operating conditions, i.e. small signal a.c. parameters.

The parameters h_{11}, h_{12}, h_{22} and h_{21} may be very easily determined either from the static characteristics of the transistor or simple open circuit and short circuit measurements.

Hence h_{11} = input resistance with the output short circuited or the slope of the input characteristic.

h_{12} = feedback ratio with the input open circuited or slope of the feedback characteristic.

h_{22} = output conductance with the input open circuited or the slope of the output characteristic.

h_{21} = current gain with the output short circuited or slope of the transfer characteristic.

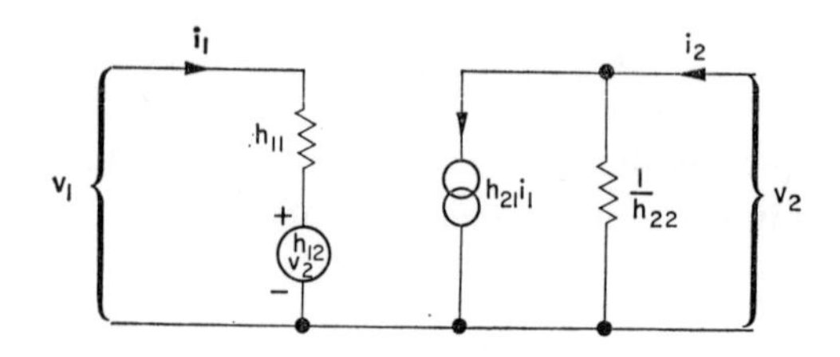

FIG. 4.3.1. Grounded base h equivalent circuit.

For other equivalent circuits it is, in general, difficult to determine the parameters by single measurements. Usually several measurements have to be made and by tedious mathematics the parameters evaluated.

A practical circuit to determine the h parameters for the common emitter connection is shown in Fig. 4.3.2. The parameters are primed (i.e. h') to show that they refer to this method of connection. In the circuit

R_1 is a high resistance, say 1 MΩ,
R_2 is a low resistance, say 100 Ω,
R_3 is a low resistance, say 10 Ω.

h'_{11}. The collector load is low enough to be considered a short circuit and with an oscillator input of v volts at (A) the voltage at (B) v_B may be noted.

Then
$$h'_{11} = \frac{v_{eb}}{i_b}$$
$$= \frac{v_B}{v/R_1}$$
$$= \frac{v_B}{v} \times R_1.$$

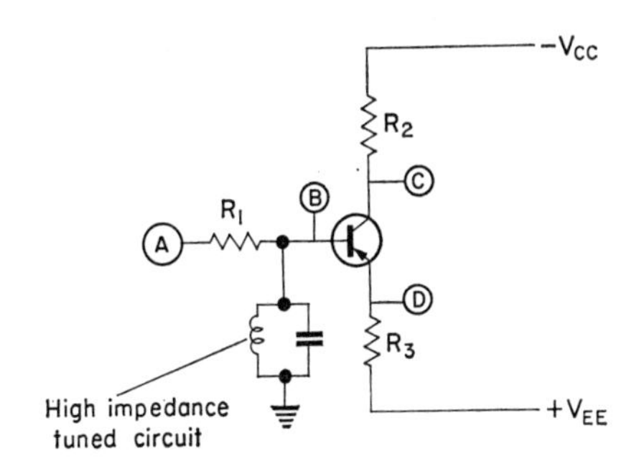

FIG. 4.3.2. Circuit to determine h parameters.

h'_{22}. With (A) open circuit and the oscillator connected to (C) the voltage at (D) v_D may be noted. Then since the input is virtually open circuited

$$h'_{22} = \frac{i_c}{v_C},$$

i.e.
$$h'_{22} = \frac{i_e}{v_C} \quad \text{since} \quad i_b \text{ is small,}$$
$$= \frac{v_D/R_3}{v_C}.$$
$$= \frac{v_D}{v_C} \times \frac{1}{R_3}.$$

h'_{12}. As in the previous case the input is virtually open circuited. Hence, if the voltage at (B) v_B is noted with an oscillator in put of v volts at the collector,

$$h'_{12} = \frac{v_B}{v}.$$

h'_{21}. In this case the oscillator is connected to (A) and the voltage measured at the point (C). Then since the collector load is small

$$h'_{21} = \frac{i_c}{i_b}$$

$$= \frac{v_C/R_2}{v/R_1}$$

$$= \frac{v_C}{v} \times \frac{R_1}{R_2}.$$

For some of these measurements very sensitive valve voltmeters or amplifiers of known gain will be required.

The hybrid parameters are real at low frequencies but complex at high frequencies. Hence to evaluate the high frequency performance the phase relationship between measured and oscillator voltages must be determined as well as their relative magnitudes.

This will prove difficult with the simple circuit shown and other techniques may have to be employed. In particular the susceptance across the output admittance and the effect of frequency on h'_{21} must be determined.

Q.4.4

Give sketches of the d.c. characteristics of *pnp* junction transistor, suitable for general purpose, low frequency applications. Typical values of the currents and voltages should be shown. Either the grounded base or grounded emitter configuration may be used.

Show how the general shape of the curves is accounted for by the physical mechanism of operation.

Sketch the equivalent circuit used to represent the behaviour of a transistor under small signal conditions. Show how the parameters of the equivalent circuit may be derived from the characteristic curves.

[B.Sc. 3, 1961]

A.4.4

Typical characteristics for a general purpose *pnp* transistor in the grounded base configuration are shown in Figs. 4.4.1, 4.4.2, 4.4.3 and 4.4.4. They may be explained by the same considerations used in A.4.1.

The input characteristic shown in Fig. 4.4.1 is essentially the forward characteristic of a juntcion diode. Although the

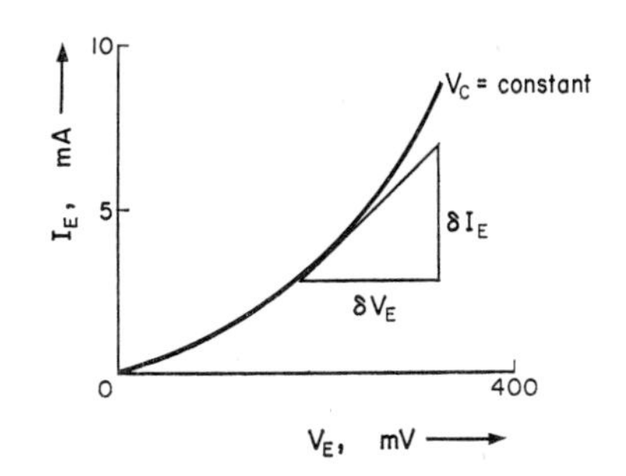

FIG. 4.4.1. Grounded base input characteristic.

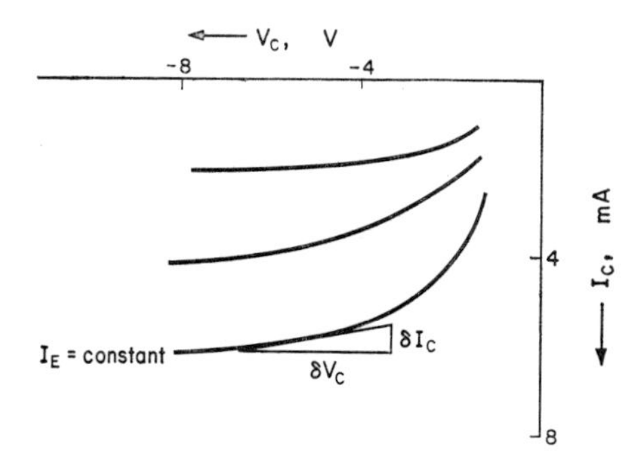

FIG. 4.4 2. Grounded base output characteristic.

collector voltage has a slight effect, the relationship is exponential in character.

The output characteristic shown in Fig. 4.4.2 is, for zero emitter current, the reverse characteristic of a *pn* junction. Any

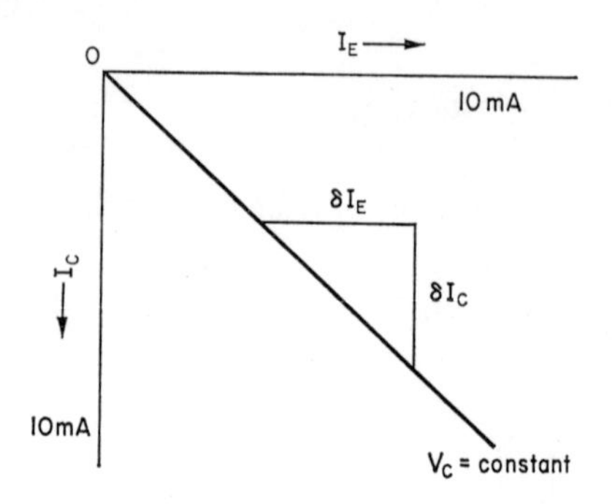

FIG. 4.4.3. Grounded base transfer characteristic.

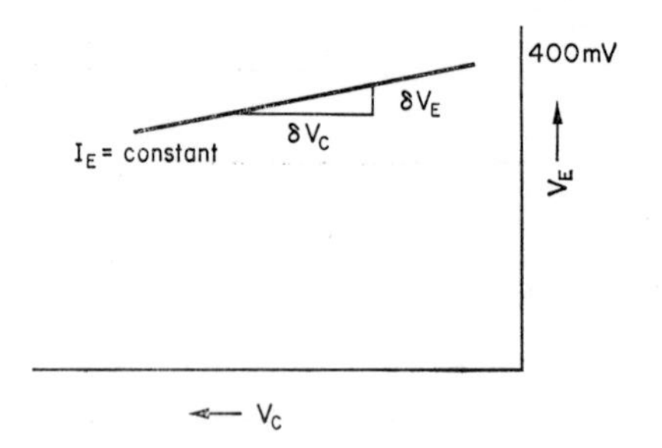

FIG. 4.4.4. Grounded base feedback characteristic.

increase in the emitter current leads to an increase in the number of minority carriers appearing at the collector junction.

Hence the output characteristics form a family of parallel curves spaced by an amount proportional to the change in emitter current.

The transfer characteristic shown in Fig. 4.4.3 is linear over the working range of the transistor. This is to be expected since to a first approximation the recombination in the base region is proportional to the current diffusing through the base region.

The feedback characteristic shown in Fig. 4.4.4 is approximately linear and depends on the variation of base width with collector voltage. If the collector voltage goes more negative the width of the depletion layer is increased, the effective base width is reduced and the recombination in the base decreased. Hence, for a given emitter, or collector current, a lower value of input voltage is required.

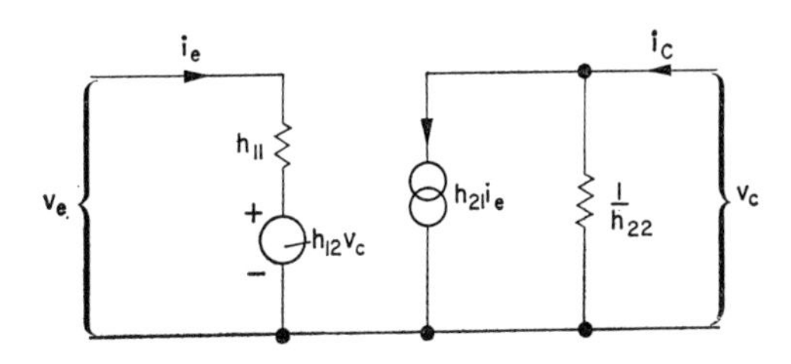

FIG. 4.4.5. Grounded base h equivalent circuit.

From the characteristics shown the hybrid parameters discussed in A.4.3 may be derived.

i.e.

$$h_{11} = \left(\frac{\delta V_E}{\delta I_E}\right) V_C = \text{constant.}$$

$$h_{22} = \left(\frac{\delta I_C}{\delta V_C}\right) I_E = \text{constant.}$$

$$h_{12} = \left(\frac{\delta V_E}{\delta V_C}\right) I_E = \text{constant.}$$

$$h_{21} = \left(\frac{\delta I_C}{\delta I_E}\right) V_C = \text{constant.}$$

The equivalent circuit is then as shown in Fig. 4.4.5.

Q.4.5

Show that the h and T parameters of a transistor are related by Table 4.1.

Evaluate when:

$$r_e = 50\ \Omega, \quad r_b = 1\ \text{k}\Omega, \quad r_c = 1\ \text{M}\Omega, \quad \alpha_0 = 0{\cdot}98$$

TABLE 4.1

Common base		Common emitter		Common collector	
h_{11}	$r_e + r_b(1-\alpha_0)$	h'_{11}	$r_b + \dfrac{r_e}{1-\alpha_0}$	h''_{11}	r_b
h_{12}	$\dfrac{r_b}{r_c}$	h'_{12}	$\dfrac{r_e}{r_c(1-\alpha_0)}$	h''_{12}	1
h_{22}	$\dfrac{1}{r_c}$	h'_{22}	$\dfrac{1}{r_c(1-\alpha_0)}$	h''_{22}	$\dfrac{1}{r_c(1-\alpha_0)}$
h_{21}	$-\alpha_0$	h'_{21}	$\dfrac{\alpha_0}{1-\alpha_0} = \beta_0$	h''_{21}	$-\dfrac{1}{1-\alpha_0} = -\gamma_0$

A.4.5

Consider the common emitter configuration. The T and h equivalent circuits are shown in Figs. 4.5.1 and 4.5.2.

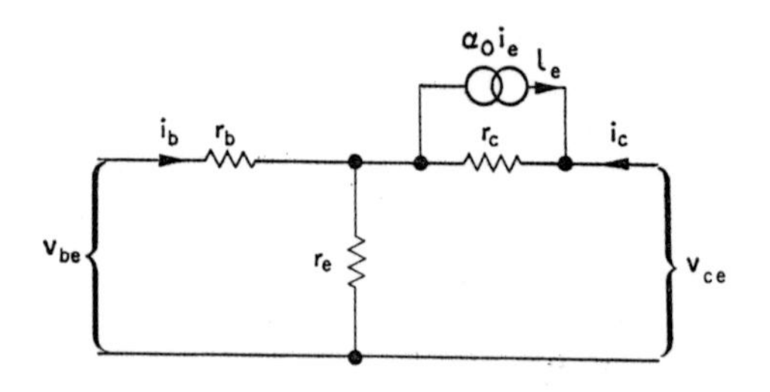

FIG. 4.5.1. Grounded emitter T equivalent circuit.

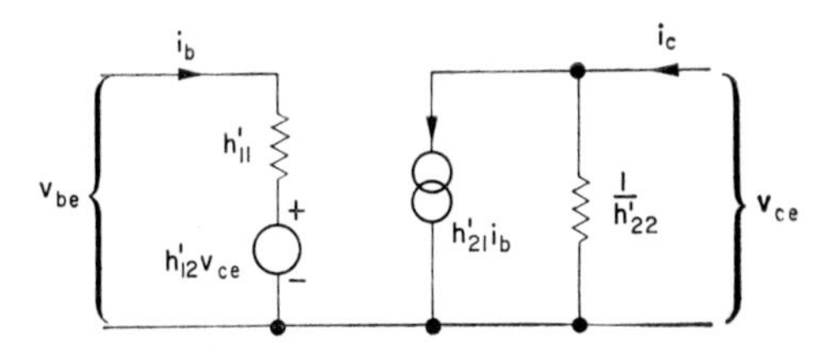

FIG. 4.5.2. Grounded emitter h equivalent circuit.

For the equivalent T circuit:

$$v_{be} = i_b(r_b+r_e)+i_c r_e, \tag{1}$$

$$\begin{aligned} v_{ce} &= i_b r_e + i_c(r_c+r_e)+\alpha_0 i_e r_c \\ &= i_b(r_e-\alpha_0 r_c)+i_c[r_e+r_c(1-\alpha_0)], \\ &\qquad \text{since } i_e = -(i_c+i_b), \\ &= -i_b\alpha_0 r_c + i_c r_c(1-\alpha_0), \\ &\qquad \text{since } r_e \ll r_c(1-\alpha_0). \end{aligned} \tag{2}$$

For the h equivalent circuit,

$$v_{be} = i_b h'_{11} + h'_{12} v_{ce},$$

$$i_c = h'_{22} v_{ce} + h'_{21} i_b.$$

Rearranging the last equation,

$$v_{ce} = -\,i_b\frac{h'_{21}}{h'_{22}} + \frac{i_c}{h'_{22}}. \tag{3}$$

Substituting for v_{ce},

$$v_{be} = i_b\left(h'_{11} - \frac{h'_{21}}{h'_{22}} \times h'_{12}\right) + i_c\frac{h'_{12}}{h'_{22}}. \tag{4}$$

Hence equating coefficients:

From (2) and (3), $$\frac{1}{h'_{22}} = r_c(1-\alpha_0)$$

or $$h'_{22} = \frac{1}{r_c(1-\alpha_0)}$$

and $$-\frac{h'_{21}}{h'_{22}} = -\alpha_0 r_c$$

or $$h'_{21} = \frac{\alpha_0}{1-\alpha_0} = \beta_0.$$

From (1) and (4), $$\frac{h'_{12}}{h'_{22}} = r_e$$

or $$h'_{12} = \frac{r_e}{r_c(1-\alpha_0)}$$

and $$h'_{11} - \frac{h'_{21}}{h'_{22}} \times h'_{12} = r_b + r_e$$

or $$h'_{11} = r_b + r_e + r_e \frac{\alpha_0}{1-\alpha_0}$$

$$= r_b + \frac{r_e}{1-\alpha_0}.$$

Substituting values:

$$h'_{22} = 50 \times 10^{-6} \text{ mho},$$
$$h'_{21} = 49,$$
$$h'_{12} = \frac{1}{400},$$
$$h'_{11} = 3500\ \Omega.$$

The relations for the common base configuration are derived in *T.D.E.*, section 4.7, and substituting values

$$h_{11} = 70\ \Omega,$$
$$h_{12} = \frac{1}{1000},$$
$$h_{22} = 10^{-6} \text{ mho},$$
$$h_{21} = -0{\cdot}98.$$

The relations for the common collector configuration may be proved by the reader and substituting for r_e, r_b, r_c and α_0

gives:

$$h''_{11} = 1000\,\Omega,$$
$$h''_{12} = 1,$$
$$h''_{22} = 50\times 10^{-6}\text{ mho},$$
$$h''_{21} = 50.$$

Note. — An alternative method for this and similar questions is given in Chapter 4 of *Signal Flow Analysis*, by Abrahams and Coverley (Pergamon Press).

Q.4.6

Show that the Z and T parameters of a transistor are related by Table 4.2.

Evaluate when:

$$r_e = 50\ \Omega,\quad r_b = 1\ \text{k}\Omega,\quad r_c = 1\ \text{M}\Omega,\quad \alpha_0 = 0{\cdot}98$$

TABLE 4.2

Common base		Common emitter		Common collector	
Z_{11}	r_e+r_b	Z'_{11}	r_e+r_b	Z''_{11}	r_c
Z_{12}	r_b	Z'_{12}	r_e	Z''_{12}	$r_c(1-\alpha_0)$
Z_{21}	$\alpha_0 r_c$	Z'_{21}	$-\alpha_0 r_c$	Z''_{21}	r_c
Z_{22}	r_c	Z'_{22}	$r_c(1-\alpha_0)$	Z''_{22}	$r_c(1-\alpha_0)$

A.4.6

Consider the common collector configuration. The T and Z equivalent circuits are shown in Figs. 4.6.1 and 4.6.2. For the T circuit

$$v_{bc} = i_b(r_b+r_c)+i_e r_c(1-\alpha_0)$$
$$= i_b r_c + i_e r_c(1-\alpha_0), \tag{1}$$

since $r_b \ll r_c$;

$$v_{ec} = i_e r_e + i_e r_c(1-\alpha_0)+i_b r_c$$
$$= i_e r_c(1-\alpha_0)+i_b r_c, \tag{2}$$

since $r_e \ll r_c(1-\alpha_0)$.

For the Z equivalent circuit

$$v_{bc} = i_b Z''_{11} + i_e Z''_{12}, \tag{3}$$
$$v_{ec} = i_e Z''_{22} + i_b Z''_{21}. \tag{4}$$

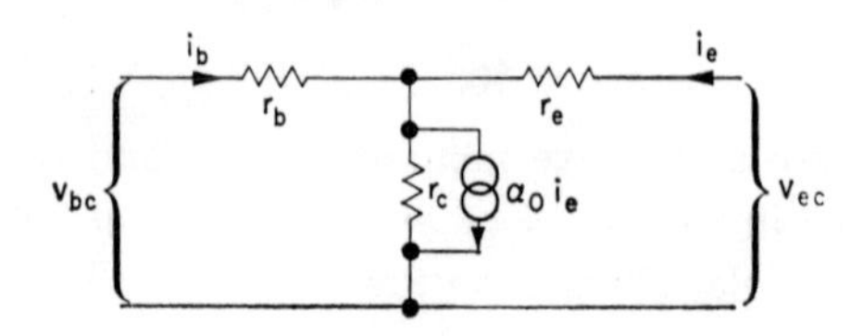

FIG. 4.6.1. Grounded collector T equivalent circuit.

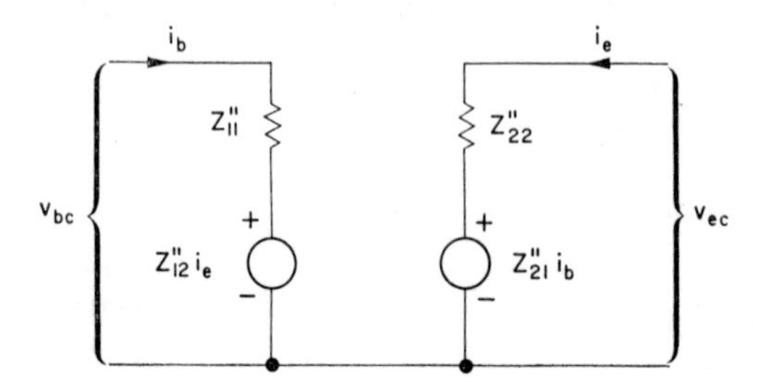

FIG. 4.6.2. Grounded collector Z equivalent circuit.

Hence, comparing coefficients:

From (1) and (3)

$$Z''_{11} = r_c,$$
$$Z''_{12} = r_c(1-\alpha_0).$$

From (2) and (4)

$$Z''_{22} = r_c(1-\alpha_0),$$
$$Z''_{21} = r_c.$$

Substituting values

$$Z''_{11} = 10^6\ \Omega,$$
$$Z''_{12} = 2\times 10^4\ \Omega,$$
$$Z''_{22} = 2\times 10^4\ \Omega,$$
$$Z''_{21} = 10^6\ \Omega.$$

The relations for the common emitter configuration are proved in *T.D.E.*, section 4.9, and substituting values

$$Z'_{11} = 1050\ \Omega,$$
$$Z'_{12} = 50\ \Omega,$$
$$Z'_{21} = -0{\cdot}98\times 10^{6}\ \Omega,$$
$$Z''_{22} = 2\times 10^{4}\ \Omega.$$

The relations for the common base configuration may be proved by the reader and substituting values

$$Z_{11} = 1050\ \Omega,$$
$$Z_{12} = 1000\ \Omega,$$
$$Z_{21} = 0{\cdot}98\times 10^{6}\ \Omega,$$
$$Z_{22} = 10^{6}\ \Omega.$$

Q.4.7

Show that the Y and T parameters of a transistor are related by Table 4.3.

Evaluate when:

$$r_e = 50\ \Omega,\quad r_b = 1\ \text{k}\Omega,\quad r_c = 1\ \text{M}\Omega,\quad \alpha_0 = 0{\cdot}98$$

TABLE 4.3

Common base		Common emitter		Common collector	
y_{11}	$\dfrac{1}{r_e+r_b(1-\alpha_0)}$	y'_{11}	$\dfrac{1-\alpha_0}{r_e+r_b(1-\alpha_0)}$	y''_{11}	$\dfrac{1}{r_b+r_e/(1-\alpha_0)}$
y_{12}	$-\dfrac{r_b}{r_c[r_e+r_b(1-\alpha_0)]}$	y'_{12}	$\dfrac{r_e}{r_c[r_e+r_b(1-\alpha_0)]}$	y''_{12}	$-\dfrac{1}{r_b+r_e/(1-\alpha_0)}$
y_{22}	$\dfrac{r_e+r_b}{r_c[r_e+r_b(1-\alpha_0)]}$	y'_{22}	$\dfrac{r_e+r_b}{r_c[r_e+r_b(1-\alpha_0)]}$	y''_{22}	$\dfrac{1}{r_b(1-\alpha_0)}$
y_{21}	$-\dfrac{\alpha_0}{r_e+r_b(1-\alpha_0)}$	y'_{21}	$\dfrac{\alpha_0}{r_e+r_b(1-\alpha_0)}$	y''_{21}	$-\dfrac{1}{r_b(1-\alpha_0)}$

A.4.7

Consider the common base circuit. The T and Y equivalent circuits are shown in Figs. 4.7.1 and 4.7.2. For the equivalent T circuit:

$$v_{eb} = i_e(r_e + r_b) + i_c r_b,$$
$$v_{cb} = i_c r_c + \alpha_0 i_e r_c + i_e r_b$$
$$= i_c r_c + i_e \alpha_0 r_c, \quad \text{since } r_b \ll \alpha_0 r_c.$$

Eliminating i_c from these equations,

$$v_{eb} = i_e(r_e + r_b) + \left(\frac{v_{cb} - i_e r_c \alpha_0}{r_c}\right) r_b$$

or

$$i_e[r_e + r_b(1-\alpha_0)] = v_{eb} - v_{cb}\frac{r_b}{r_c}$$

or

$$i_e = \frac{v_{eb}}{r_e + r_b(1-\alpha_0)} - \frac{r_b v_{cb}}{r_c[r_e + r_b(1-\alpha_0)]}. \tag{1}$$

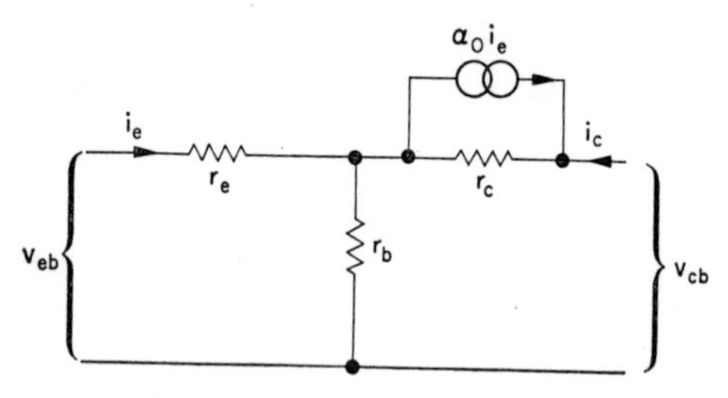

FIG. 4.7.1. Grounded base T equivalent circuit.

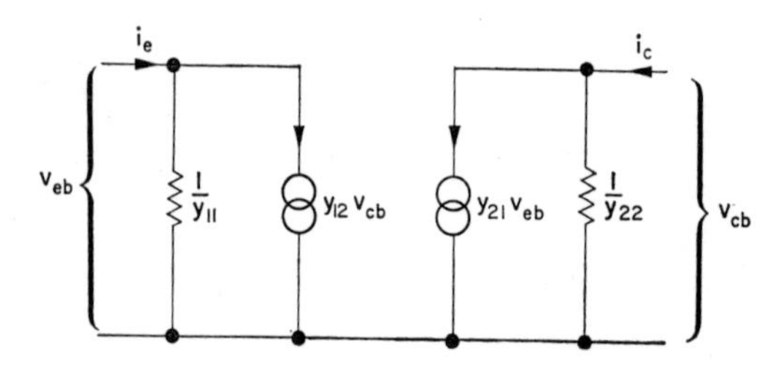

FIG. 4.7.2. Grounded base Y equivalent circuit.

Eliminating i_e,

$$v_{eb} = \left(\frac{v_{cb}-i_c r_c}{\alpha_0 r_c}\right)(r_e+r_b)+i_c r_b$$

or

$$i_c[r_e+r_b(1-\alpha_0)] = \left(\frac{r_e+r_b}{r_c}\right)v_{cb}-\alpha_0 v_{eb}$$

or

$$i_c = \frac{(r_e+r_b)v_{cb}}{r_c[r_e+r_b(1-\alpha_0)]}-\frac{\alpha_0 v_{eb}}{[r_e+r_b(1-\alpha_0)]}. \tag{2}$$

For the Y equivalent circuit,

$$i_e = v_{eb}y_{11}+v_{cb}y_{12}, \tag{3}$$

$$i_c = v_{cb}y_{22}+v_{eb}y_{21}. \tag{4}$$

Hence, comparing coefficients:

From (1) and (3),

$$y_{11} = \frac{1}{r_e+r_b(1-\alpha_0)},$$

$$y_{12} = -\frac{r_b}{r_c[r_e+r_b(1-\alpha_0)]}.$$

From (2) and (4),

$$y_{22} = \frac{r_e+r_b}{r_c[r_e+r_b(1-\alpha_0)]},$$

$$y_{21} = -\frac{\alpha_0}{r_e+r_b(1-\alpha_0)}.$$

Substituting values:

$$y_{11} = \frac{1}{70}\text{ mho},$$

$$y_{12} = -\frac{1}{70000}\text{ mho},$$

$$y_{22} = \frac{3}{200000}\text{ mho},$$

$$y_{21} = -\frac{7}{500}\text{ mho}.$$

The relations for the common collector configuration are derived in *T.D.E.*, section 4.1.1, and substituting values

$$y''_{11} = \frac{1}{3500} \text{ mho},$$

$$y''_{12} = -\frac{1}{3500} \text{ mho},$$

$$y''_{22} = 0{\cdot}05 \text{ mho},$$

$$y''_{21} = -0{\cdot}05 \text{ mho}.$$

The relations for the common emitter configuration may be proved by the reader and substituting values

$$y'_{11} = \frac{1}{3500} \text{ mho},$$

$$y'_{12} = \frac{1}{1200000} \text{ mho},$$

$$y'_{22} = \frac{3}{200000} \text{ mho},$$

$$y'_{21} = \frac{7}{500} \text{ mho}.$$

CHAPTER 5

Rectifiers and Stabilizers

Q.5.1

Draw the circuit diagram of a simple d.c. stabilizer employing one resistor and a Zener diode. Derive expressions for the stabilization ratio and output resistance, in terms of the component parameters assuming a d.c. source of zero internal impedance.

Such a stabilizer circuit has a ballast resistor of 80 Ω with a diode having a Zener voltage of -9 V and a constant slope resistance of 4 Ω. What is the change in output voltage if the load current falls from 50 mA to 20 mA?

[H.N.D. 3, 1964]

A.5.1

The circuit of a simple d.c. stabilizer, with one Zener diode and a resistor, is given in Fig. 5.1.1. The currents and voltages are indicated by appropriate symbols. The Zener diode may be assumed to have a reverse breakdown voltage of V_Z and a constant slope resistance, when conducting beyond reverse breakdown, of $R_Z \Omega$.

Then for an output of V_0 the current flowing through the Zener diode is given by

$$I_Z = \frac{V_0 - V_Z}{R_Z}$$

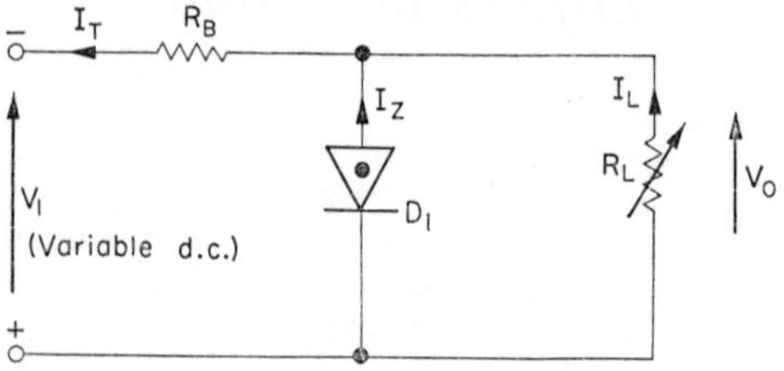

FIG. 5.1.1. Simple Zener diode stabilizer.

At the same time, for a load current of I_L, the total current drawn from the supply is given by $I_T = I_L + I_Z$ and the input voltage is $V_1 = I_T R_B + V_0$.

Therefore

$$V_1 = R_B(I_L + I_Z) + V_0$$

or

$$V_1 = I_L R_B + (V_0 - V_z)\frac{R_B}{R_Z} + V_0,$$

where

$$I_L = \frac{V_0}{R_L}$$

and hence

$$V_1 = V_0\left(1 + \frac{R_B}{R_Z} + \frac{R_B}{R_L}\right) - V_Z\left(\frac{R_B}{R_Z}\right).$$

To calculate the approximate effect of a change in the input voltage we may assume that V_Z is constant and that $R_Z \ll R_L$ and therefore R_B/R_L may be ignored in comparison with R_B/R_Z.

Then $$\delta V_1 \simeq \delta V_0\left(1+\frac{R_B}{R_Z}\right)$$

considering small changes of voltage

or $$\frac{\delta V_0}{\delta V_1}=\frac{R_Z}{R_B+R_Z},$$

voltage stability factor S.

Or, if we wish to find the effect of a variation in the load current, assuming that the source supplies a constant voltage V, then by rearrangement of the previous equation for V_1,

$$V_0\left(1+\frac{R_B}{R_Z}\right)=V_Z\left(\frac{R_B}{R_Z}\right)-I_L R_B+V_1.$$

Hence $$\delta V_0\left(1+\frac{R_B}{R_Z}\right)=-\ \delta I_L R_B$$

or $$\frac{\delta V_0}{\delta I_L}=-\frac{R_B R_Z}{R_B+R_Z}\,\Omega=\text{ output resistance.}$$

Substituting the values given for R_B and R_Z in this example,

where $$\delta I_L=-30\times 10^{-3}\text{ A},$$

we have that $$\delta V_0=(-30\times 10^{-3})\left(\frac{-80\times 4}{80+4}\right)\text{ V}$$

or $$\delta V_0=\frac{3\times 3{\cdot}2}{84}\simeq 0{\cdot}11\text{ V}.$$

That is the output voltage will rise by about 110 mV for a fall in load current of 30 mA.

Q.5.2

Why are d.c. stabilizer circuits using gas-filled valves not convenient for low-voltage applications?

Sketch the circuit of a simple stabilizer using a silicon junction reference diode and a power transistor as the series element. Explain the function of each component and indicate suitable component values for a circuit to supply 1 A at 6 V from a nominally 12 V supply, which may vary by ± 2 V.

[H.N.C. 3, 1963]

A.5.2

The gas-filled diodes employed as voltage stabilizers have striking voltages of 60 V or more. In addition, the striking voltage is always a few volts more than the maintaining voltage.

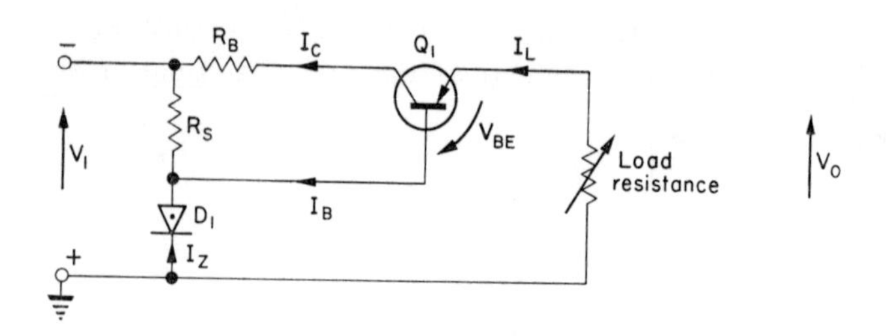

FIG. 5.2.1. Series transistor stabilizer circuit.

For these reasons it is not convenient to use a gas diode in a low-voltage stabilizer intended for transistor circuit. In contrast the silicon junction reference diode can be manufactured with a reverse breakdown voltage in the range from -2 V to greater than -100 V, and does not require a striking voltage in excess of the maintaining voltage.

The circuit of a series stabilizer is drawn as Fig. 5.2.1, incorporating a power transistor Q_1, a Zener diode D_1 and two resistors R_B and R_S.

The transistor in this stabilizer circuit acts as a current divider, since the ratio of emitter current to base current is given approximately by the common emitter current gain β, where the emitter current passes through the load. Any change of current through the diode is thus a small fraction of a change in the load current. Under this condition the voltage across the diode will be held appreciably constant, thus stabilizing the voltage across the load.

The shunt resistor R_S is chosen to give a suitable current through the Zener diode. This current should be on the straight line portion of the diode characteristic (e.g. between 10–50 mA for the typical small Zener diode).

The series "ballast" resistor R_B is employed to minimize the power dissipation in the series transistor. This resistance of R_B is chosen to give a volt drop across R_B which is slightly less than the difference between minimum input and maximum output voltages, at the maximum value of load current.

For the calculation we may assume that the load current is steady at 1 A at a voltage of 6 V. The minimum input voltage is 10 V and a minimum volt drop of 1 V should be allowed across the transistor.

Hence the ballast resistor is given by

$$R_B = \frac{10-1-6}{1} = 3\ \Omega$$

and the dissipation in R_B is 3 W.

The maximum voltage expected across the transistor is thus $14-3-6 = 5$ V, and the transistor dissipation is then about 5 W. A common emitter current gain of at least 25 may be assumed for an emitter current of 1 A. Then the base current is 40 mA. Also a minimum current of 10 mA should flow in the Zener diode, to avoid the curved portion of the characteristic. If the base to emitter voltage of the transistor is 300 mV at 1 A, then the Zener diode voltage should be 6·3 V (at 10 mA). Under the minimum input voltage condition the volt drop across R_S is $10-6{\cdot}3 = 3{\cdot}7$ V with a current of 50 mA. Thus R_S should be about $3{\cdot}7/50\times10^3\ \Omega$, and this suggests a preferred resistance value of 68 Ω for R_S.

With the maximum input of 14 V and assuming a diode slope resistance of 2 Ω, the current through R_S increases by

4000/70 mA. This means that the maximum current through the Zener diode is nearly 70 mA and the maximum power dissipation in R_S is 0·7 W. Also the maximum power dissipation in the Zener diode is approximately 0·4 W.

To summarize these results:

$R_B = \ \ 3\ \Omega$ and 3 W,

$R_S = 68\ \Omega$ and 1 W,

Q_1 must dissipate 5 W at 1 A, with $\beta_{min} = 25$,

D_1 has $V_Z = -6{\cdot}3$ V at 10 mA, with a maximum current rating of at least 70 mA, a slope resistance of 2 Ω and dissipation of $\frac{1}{2}$ W.

Q.5.3

Explain the operation of a germanium *pn* junction diode. Sketch a typical current/voltage characteristic.

A diode has a characteristic which may be assumed to be a constant slope resistance of 100 Ω passing through the point $I = 0$, $V = 0{\cdot}5$ V in the conducting direction and a constant resistance of 10 kΩ in the reverse direction. The diode is connected in series with a sinusoidal e.m.f. of peak value 4 V and a load resistance of 500 Ω. Sketch the current waveform and calculate the mean current. Comment on the validity of the assumed characteristic in practice.

[I.E.E., May 1962]

A.5.3

The answer to the first part of this question is covered in A.2.1.

Figures 5.3.1 and 5.3.2 show a typical diode characteristic and that of the diode used in this question. The latter characteristic enables the equivalent circuit shown in Fig. 5.3.3 to be drawn. In this circuit the diodes are ideal rectifiers.

Voltage and current waveforms are shown in Fig. 5.3.4, the peak forward and reverse currents and the limits of forward current flow being obtained as follows.

$$\text{Peak forward current} = \frac{4-0{\cdot}5}{600} \times 1000 = 5{\cdot}83 \text{ mA}$$

$$\text{Peak reverse current} = \frac{4}{10{\cdot}5 \times 10^3} = 0{\cdot}38 \text{ mA}$$

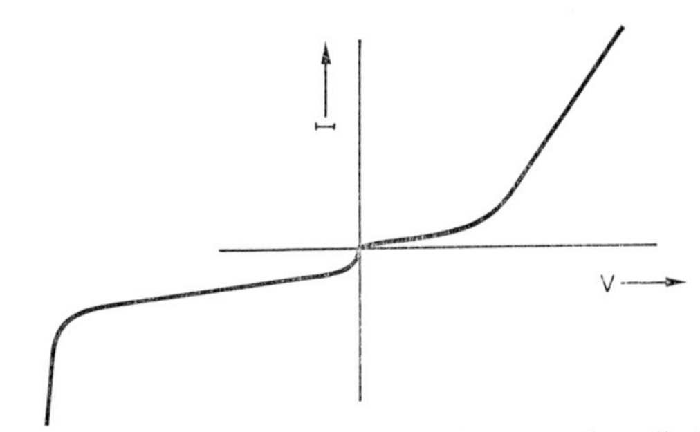

FIG. 5.3.1. Typical characteristics of a germanium diode.

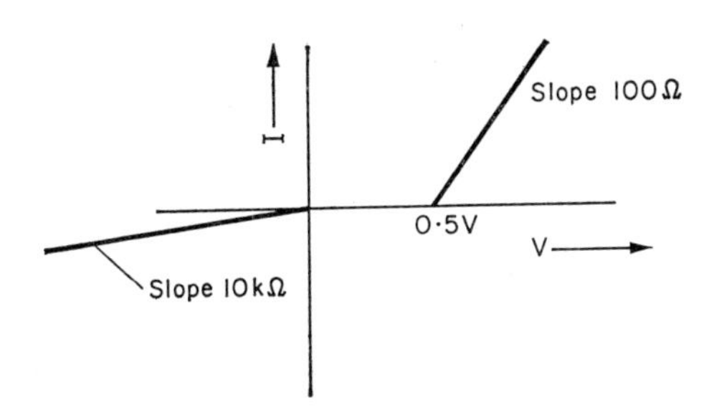

FIG. 5.3.2. Assumed diode curve for Q.5.3.

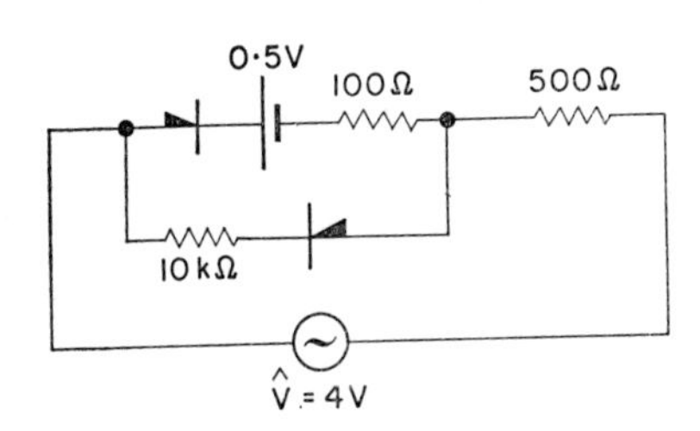

FIG. 5.3.3. Circuit of Q.5.3. redrawn with "perfect" diodes.

Limits of forward current flow

$$= \sin^{-1}\frac{0{\cdot}5}{4} \quad \text{and} \quad \pi - \sin^{-1}\frac{0{\cdot}5}{4}$$

$$= 7{\cdot}2^\circ \quad \text{or} \quad 0{\cdot}126 \text{ rad and } 172{\cdot}8^\circ \text{ or } 3{\cdot}02 \text{ rad.}$$

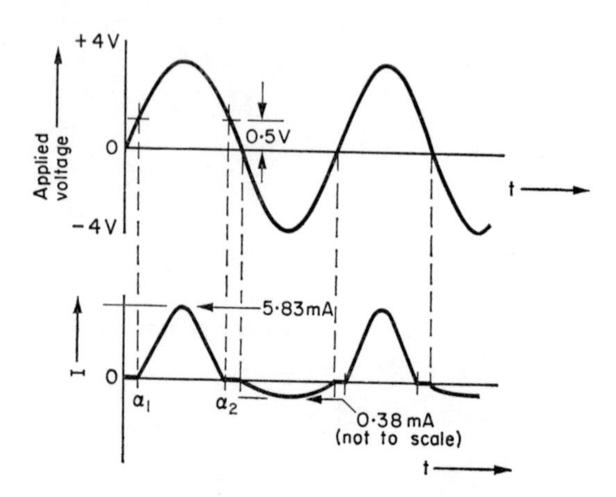

FIG. 5.3.4. Voltage and current waveforms for A.5.3.

The mean current is then given by

$$I_{\text{mean}} = \frac{1}{2\pi}\left[\int_{\alpha_1}^{\alpha_2}\left(\frac{4\sin\omega t - 0{\cdot}5}{600}\right)d(\omega t) + \int_{\pi}^{2\pi}\frac{4\sin\omega t}{10500}\,d(\omega t)\right]\text{A}$$

$$= \frac{1}{2\pi}\left[\frac{20}{3}\Big(-\cos\omega t\Big)_{\alpha_1}^{\alpha_2} - \frac{5}{6}\Big(\omega t\Big)_{\alpha_1}^{\alpha_2} + \frac{8}{21}\Big(-\cos\omega t\Big)_{\pi}^{2\pi}\right]\text{mA}$$

$$= \frac{1}{2\pi}\left[\frac{20}{3}(2\times 0{\cdot}9921) - \frac{5}{6}\times 2{\cdot}89 + \frac{8}{21}(-2)\right]$$

$$= \frac{1}{2\pi}(13{\cdot}23 - 2{\cdot}41 - 0{\cdot}76)$$

$$= 1{\cdot}6 \text{ mA.}$$

Examination of Figs. 5.3.3 and 5.3.4 shows that the assumed characteristic is valid provided that

(1) The applied voltage is several times the diode barrier potential (i.e. 0·5 V in this case).
(2) The reverse resistance is very much higher than the load resistance.
(3) The diode is not operated under breakdown conditions.

Q.5.4

Sketch a typical current–voltage characteristic for a germanium junction rectifier and describe briefly its features.

A 16 V r.m.s. 50 c/s supply is connected in series with a germanium junction rectifier and a limiting resistor to charge a 12 V battery. Assuming a voltage drop of 0·6 V across the rectifier independent of the forward current and negligible reverse current calculate the peak current and the power dissipated in the rectifier for a mean charging current of 1 A.

[I.E.E., Nov. 1962]

A.5.4

The answer to the first part of this question is covered (in more detail than is required here) in A.2.1.

The numerical solution is similar to the previous question, but in this case current flows when the applied voltage is greater than 12·6 V.

i.e. the limits of conduction are $\sin^{-1}\frac{12{\cdot}6}{16{\cdot}2}$ and $-\sin^{1}\frac{12{\cdot}6}{16{\cdot}2}$

i.e. 33·9° or 0·592 rads and 146·1° or 2·55 rads.

The mean current is then given by

$$I_{\text{mean}} = \frac{1}{2\pi}\int_{0\cdot 592}^{2\cdot 55} \frac{16\sqrt{2}\sin\omega t - 12{\cdot}6}{R}\, d(\omega t)$$

$$= \frac{22{\cdot}6}{2\pi R}\left(-\cos\omega t - 0{\cdot}5580\,\omega t\right)_{0\cdot 592}^{2\cdot 55}$$

$$= \frac{22{\cdot}6}{2\pi R}(0{\cdot}83 - 1{\cdot}42 + 0{\cdot}83 + 0{\cdot}33) = \frac{12{\cdot}9}{2\pi R}.$$

Therefore

$$R = \frac{12{\cdot}9}{2\pi I_{\text{mean}}}$$

$$= 2{\cdot}05\ \Omega.$$

Then

$$I_{\text{peak}} = \frac{22{\cdot}6-12{\cdot}6}{2{\cdot}05}$$

$$= \underline{4{\cdot}87\ \text{A}.}$$

The power dissipated in the rectifier

$$= \frac{1}{\tau}\int_{t_1}^{t_2} vi\ dt = \frac{1}{2\pi}\int_{0{\cdot}592}^{2{\cdot}55} vi\ d(\omega t)\quad \text{W},$$

where v = instantaneous voltage = 0·6 V,

i = instantaneous current $= \dfrac{16\sqrt{2}\sin t-12{\cdot}6}{2{\cdot}05}$,

τ = period = 0·02 sec,

i.e. power $= 0{\cdot}6\times 1$

$$= \underline{0{\cdot}6\ \text{W}.}$$

Q.5.5

A 20 V d.c. supply of negligible output impedance is connected through a 470 Ω resistor to a Zener diode in shunt with a 1 kΩ load resistor which carries a current of 6 mA. A 5% change in supply voltage causes a 0·1% change in load voltage. Calculate the slope resistance and reverse voltage breakdown of the diode. The slope resistance of the diode may be assumed to be independent of current.

[I.E.E., June 1963]

A.5.5

The circuit described in this question is reproduced in Fig. 5.5.1.

The voltage across the load is clearly 6 V (i.e. 6 mA through 1000 Ω) and a 0·1% change in load voltage is thus 6 mV.

Using the expression derived in A.5.1, we have that

$$\frac{\delta V_0}{\delta V_1} = \frac{R_Z}{R_B+R_Z},$$

where $\delta V_1 = 1$ V (i.e. 5% of 20 V).

Therefore $$R_Z = \left(\frac{6}{1000}\right)\times(470+R_Z)$$

or $$1000\ R_Z = 6\times 470-6\ R_Z.$$

Hence $$R_Z = \frac{6\times 470}{994} = 2{\cdot}85\ \Omega \text{ (say } 2{\cdot}9\ \Omega).$$

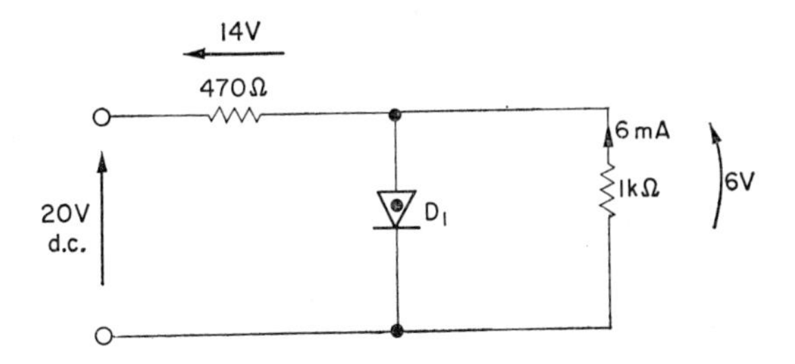

FIG. 5.5.1. Stabilizer circuit for A.5.5.

The volt drop accross the series ballast resistor (of 470 Ω) is 14 V, and so the total current drawn from the supply is

$$I_T = \frac{14}{470}\ \text{A} = 30\ \text{mA},$$

of which 24 mA flows through the Zener diode. The reverse breakdown voltage of this diode is therefore given by

$$V_Z = 6-(24\times 10^{-3}\times 2{\cdot}85) = 5{\cdot}93\ \text{V}.$$

Q.5.6

Explain the mechanism of operation of a junction diode and account for the following features:

(a) The non-linear current–voltage characteristic.

(b) The variation of the junction capacitance, with the reverse bias voltage.

(c) The occurrence of breakdown at a well-defined reverse bias voltage.

A junction diode has a current–voltage characteristic which can be approximated by assuming a forward resistance of 10 Ω and a reverse resistance of 10 kΩ provided that the reverse voltage does not exceed the breakdown value of 180 V. The diode is connected in series with a 500 Ω resistor and a generator having a sinusoidal terminal voltage of V volts, r.m.s. What is the mean value of the voltage across the 500 Ω resistor, assuming that V is too small to cause breakdown? For what value of V will breakdown occur?

[B.Sc. 3, 1962]

A.5.6

In this solution a simpler approach than that involving energy level diagrams has been given.

Under static condition diffusion of electrons and holes gives rise to a depletion region at a *pn* junction. This is a layer where holes and electrons recombine to produce a region of intrinsic semiconductor material with a potential barrier across it. Any current flow across the junction, due to thermally produced minority carriers, is immediately balanced by the flow of majority carriers.

When a *pn* junction is forward biased the potential difference across the depletion layer is reduced and the width of the layer itself is reduced. Thus, a greater number of majority carriers, i.e. free electrons from the *n*-type region and holes from the *p*-type region, will succeed in overcoming the potential barrier and the forward current increases. The reverse current, however, is unaltered since it depends on the number of minority carriers and hence on the temperature. A small forward voltage will therefore result in a forward current which is much greater than the reverse current.

If the polarity of the applied voltage is reversed the potential barrier at the junction is increased and the width of the depletion layer is increased. This reduces the forward current while the reverse current is as before, unaffected. As the reverse bias is increased the forward current becomes negligibly small and the total current approaches the constant value of the reverse current.

Thus, a *pn* junction allows a large current to flow in one direction but only a very small current in the other direction.

(a) The non-linear current–voltage characteristic follows from the rectifying action of the *pn* junction. It has a low forward resistance and a high reverse resistance so that there is a change in the slope of the characteristic at the origin.

(b) When the voltage across the *pn* junction changes, the width of the depletion layer also changes and a charging or discharging current must flow. Thus there is in effect a capacitance between the *p*-type and *n*-type regions and since the width of the depletion layer increases with increase of reverse bias the junction capacitance will decrease.

(c) The breakdown that occurs at a well defined reverse voltage is the Zener breakdown. This is due to high electric field developed across the depletion region being of sufficient magnitude to break to covalent bonds. The hole–electron pair produced give rise to high currents associated with breakdown. In many cases these electrons are accelerated and produce further ionization, i.e. avalanche breakdown.

For the positive half-cycle:

Peak value of potential difference across

$$\text{the } 500\ \Omega \text{ resistor} = \frac{500}{510}\sqrt{2}\ V.$$

Therefore

Mean value of the potential difference

$$\text{taken over the whole cycle} = \frac{50}{51}\,\frac{\sqrt{2}\ V}{\pi}\ V.$$

For the negative half-cycle:

Peak value of the potential difference across

$$\text{the } 500\ \Omega \text{ resistor} = \frac{500}{10500}\sqrt{2}\ V.$$

Therefore

$$\text{Mean value over 1 cycle} = \frac{\sqrt{2}\ V}{\pi}\left(\frac{50}{51}\ \frac{-5}{105}\right)$$

$$= 0{\cdot}45\ V\ (0{\cdot}981 - 0{\cdot}0486)$$

$$= \underline{0{\cdot}374\ V}$$

Breakdown will occur when

$$180 = \frac{10000}{10500} \times \sqrt{2}\ V.$$

Therefore $V = \dfrac{180 \times 105}{\sqrt{2} \times 100} = \underline{133{\cdot}7\ \text{V. r.m.s.}}$

Q.5.7

Write a short note justifying linear approximations for semiconductor diode characteristics.

The input to the circuit shown in Fig. 5.7.1 is a 1 kc/s square wave of 40 V peak-to-peak amplitude. Sketch to scale the output voltage if the diode has a forward resistance of 100 Ω and an infinite reverse resistance.

What would be the output voltage waveform if the diode were replaced by one which exhibited Zener breakdown at 10 V? [I.E.E., Nov. 1963]

A.5.7

The static characteristic of a semiconductor diode follows an exponential law. For the forward direction the initial rate of increase of current with voltage is low but above a certain value increases very rapidly. This rapid increase may be compared to a straight line and the characteristic approximated to

$$I = G(V-V_0)$$

where G corresponds to the conductance of the diode and V_0 the starting voltage.

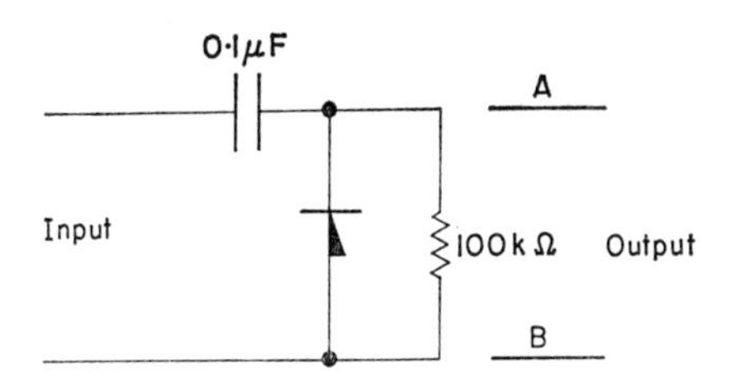

FIG. 5.7.1. Diode circuit for Q.5.7.

With a certain loss of accuracy the constant V_0 may be omitted.

If we consider the dynamic relation, the characteristic is more linear and approaches a straight line as the load resistance is increased.

For the reverse direction the ratio of voltage to current is very high and normally we may assume the reverse characteristic lies along the voltage axis. This assumes that breakdown conditions are not reached, when the reverse characteristic is

$$I = G_1(V-V_B),$$

V_B being the breakdown voltage.

The output waveform depends on the time constant for the circuit.

i.e. in the positive direction (A positive with respect to B)

$$\text{time constant} = 0{\cdot}1 \times 10^{-6} \times 10^{5} = 10 \text{ msec}$$

in the negative direction (A negative with respect to B)

$$\text{time constant} = 0{\cdot}1 \times 10^{-6} \times 10^{2} = 10\ \mu\text{sec.}$$

Since the period is 1 msec the capacitor, assuming negligible source impedance, will only charge up slightly with A positive but will be fully charged with B positive with respect to A.

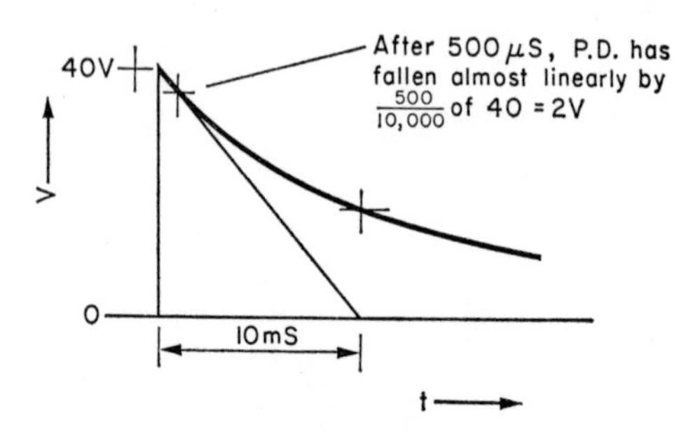

FIG. 5.7.2. Discharge curve in positive half-cycle.

Hence, as the input rises from -20 V to $+20$ V, A initially at earth potential will rise by $+40$ V. The capacitor will then charge slowly, the output falling by 2 V during the half-cycle as shown in Fig. 5.7.2.

The input and output voltages then fall by 40 V, giving an output waveform as shown in Fig. 5.7.3. The low time constant when B is positive with respect to A enables the capacitor to charge up completely in the opposite polarity.

When the diode is replaced by the Zener diode, assumed to have a forward resistance of the same order as before, the waveform will be as shown in Fig. 5.7.4. The maximum positive voltage of A is clamped at $+10$ V and during the half-cycle falls by 1/20th, i.e. 0·5 V.

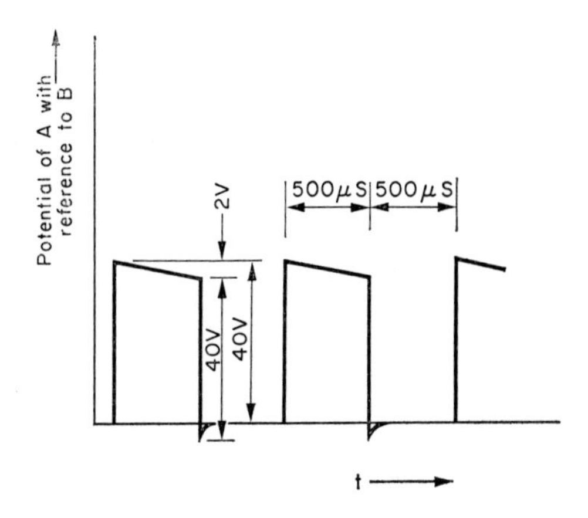

FIG. 5.7.3. Output waveform with normal diode in circuit 5.8.

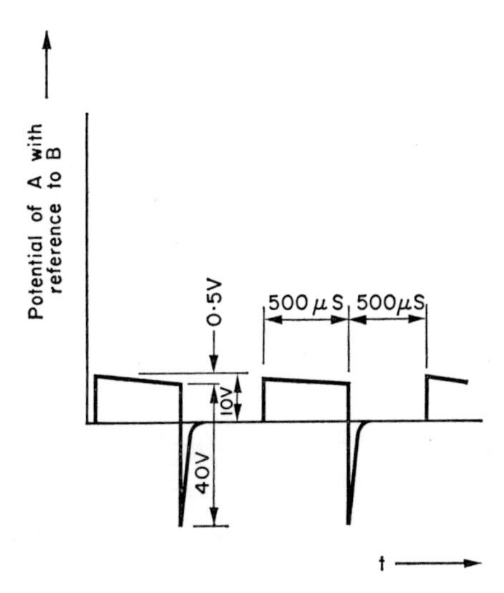

FIG. 5.7.4. Output waveform with Zener diode in circuit 5.8.

Q.5.8

The two diodes in Fig. 5.8.1 have zero forward resistance and infinite reverse resistance and the 6 V supplies have negligible resistance. The e.m.f. of the generator is represented by $10 \sin 2\pi \times 10^4\, t$ volts. Deduce expressions for the voltage V_1 and sketch the waveform.

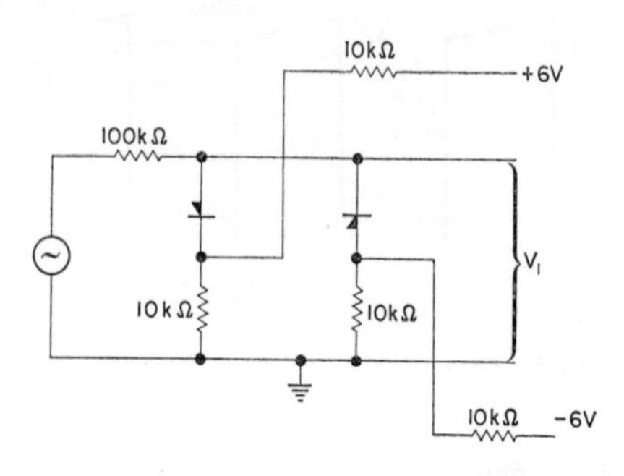

FIG. 5.8.1. Circuit, with two diodes, for Q.5.8.

A voltage of this waveform is subsequently applied to a differentiating circuit where the relationship between the output voltage V_2 and the input voltage V_1 is given by

$$V_2 = 10^{-5} \times \frac{dV_1}{dt}.$$

Sketch the waveform of the output voltage and calculate the maximum value.

[B.Sc. 3, 1962]

A.5.8

The voltage V_1 will be clipped at $+3$ V and -3 V, so that the waveform will be as shown in Fig. 5.8.2.

The diode will conduct in the half-cycle at a value of t given by

$$3 = 10 \sin 2\pi \times 10^4 \times t.$$

Therefore
$$\sin^{-1} 0{\cdot}3 = 2\pi \times 10^4 \times t,$$
$$17{\cdot}46 \times \frac{\pi}{180} = 2\pi\ 10^4 t,$$
$$t = \frac{17{\cdot}46 \times 10^6}{2 \times 180 \times 10^4}\ \mu\text{sec},$$
$$= 4{\cdot}85\ \mu\text{sec}.$$

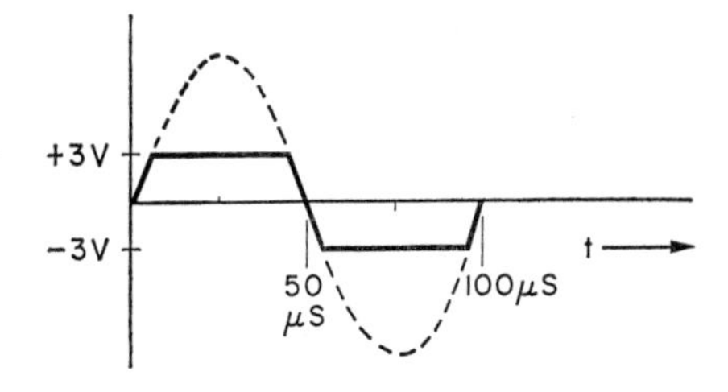

FIG. 5.8.2. Usual voltage output from circuit 5.12.

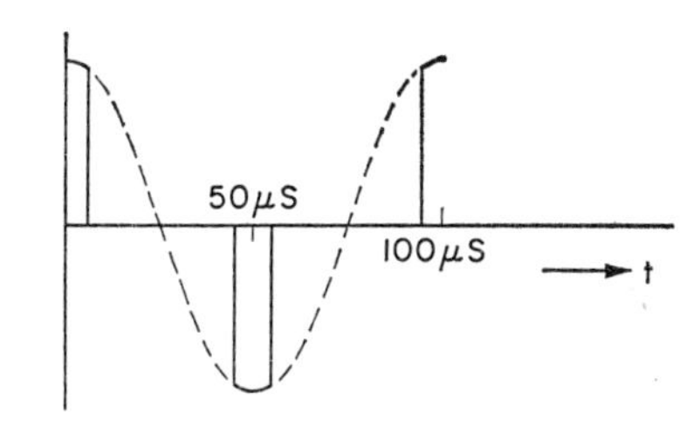

FIG. 5.8.3. Differentiated output from circuit 5.12.

Therefore $V_1 = 10 \sin 2\pi\ 10^4\ t$ volts for
$$0 \leqslant t \leqslant 4{\cdot}85\ \mu\text{sec},$$
$$45{\cdot}15\ \mu\text{sec} \leqslant t \leqslant 54{\cdot}85\ \mu\text{sec},$$
$$95{\cdot}15\ \mu\text{sec} \leqslant t \leqslant 100\ \mu\text{sec},$$
and $V_1 = 3$ V for
$$4{\cdot}85\ \mu\text{sec} \leqslant t \leqslant 45{\cdot}15\ \mu\text{sec},$$
$$54{\cdot}85\ \mu\text{sec} \leqslant t \leqslant 95{\cdot}15\ \mu\text{sec}.$$

When differentiated the waveform will be as shown in Fig. 5.8.3. The maximum value of V_2 is when $t = 0$, $t = 50\ \mu\text{sec}$, $t = 100\ \mu\text{sec}$, i.e. when $V = 10 \sin 2\pi\, 10^4\, t$. Therefore

$$V_2 = 10^{-5} \times \frac{dV_1}{dt} = 10^{-5} \times 10 \cdot 2\pi\, 10^4 \cos 2\pi\, 10^4 \times t$$

$$= 2\pi \cos 2\pi\, 10^4\, t \text{ volts.}$$

Therefore the maximum value of V_2 is 2π volts.

CHAPTER 6

Voltage Amplifiers

Q.6.1

Draw the circuit diagram of a transistorized a.f. amplifier, using the common emitter mode of operation.

Describe how the necessary biasing conditions are obtained and explain how both current and voltage amplification may be obtained using such an amplifier.

[H.N.D. 3, 1961]

A.6.1

The circuit diagram of the amplifier is shown in Fig. 6.1.1. The standing d.c. potential between the base and emitter V_{BE}, and hence the base current I_B, is decided by the potential divider R_1, R_2 and the emitter resistor R_E. The value of the base current I_B determines the collector current I_C, the latter being roughly independent of the collector voltage V_C as long as the collector voltage is greater than the knee voltage. The value of

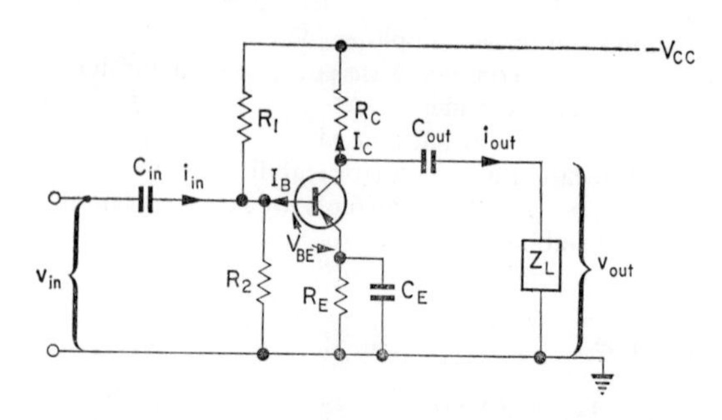

FIG. 6.1.1. Circuit diagram of a.f. amplifier.

the collector resistor R_C and the supply voltage $-V_{CC}$ then decides the actual collector voltage.

The arrangement to provide base bias is important since it should provide stabilization of the mean collector current. This is particularly important in common emitter amplifiers (see A.6.5). In the circuit shown the base voltage V_B is fixed by the potential divider and any tendency for the collector current to increase, increases the voltage drop across the emitter resistor R_E. This in turn lowers the base emitter voltage V_{BE} and hence the base current I_B. A reduction in base current I_B then reduces the collector current I_C and tends to counteract the original change.

As an a.f. amplifier the capacitors C_{in}, C_{out} and C_E should have negligible reactance at the operating frequency and the potential divider resistances R_1 R_2 should be much greater than the input impedance of the transistor. A compromise must be made here between the conflicting requirements of d.c. base bias conditions and a.c. input impedance.

If an alternating current i_{in} is superimposed on the d.c. base current the charge in the base region is modified. This produces a much larger variation in the current flowing from the emitter to the collector and the current amplification β between base and collector may be fifty times. The overall current gain between input and output is somewhat less then this due to losses in the bias resistors and the division of current between the collector resistor R_C and the load impedance Z_L. The latter is often the input impedance of a following stage of amplification.

The action of a transistor is essentially that of a current amplifier but it is easily connected as a voltage amplifier. Due to the relatively low input impedance R_{in} the input voltage v_{in} required for a current change i_{in} is low, while the high output impedance enables an appreciable output variation v_{out} to be obtained. In the case of a voltage amplifier the load impedance Z_L should be as high as possible and to a first approximation the voltage gain is given by

$$-\beta \frac{R_c}{R_{in}}.$$

The negative sign indicates an output voltage in antiphase to the input voltage.

Q.6.2

Compare qualitatively common base and common emitter transistor amplifier stages in respect of (a) input and output resistances, (b) cut-off frequency, and (c) suitability for being cascaded.

Explain in general terms how the differences arise.

Explain what is meant by the term "alpha cut-off frequency".

[C. and G. 5, 1962]

A.6.2

(a) Simple forms of common base and common emitter amplifiers are shown in Figs. 6.2.1 and 6.2.2. The input resistances depend on the transistor characteristics and to a lesser

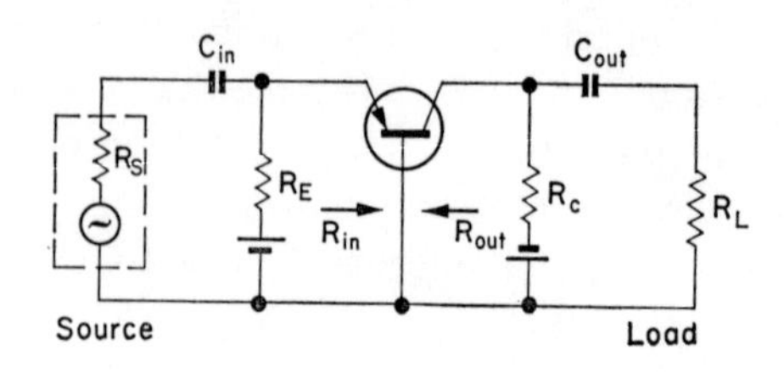

FIG. 6.2.1. Simple common base amplifier.

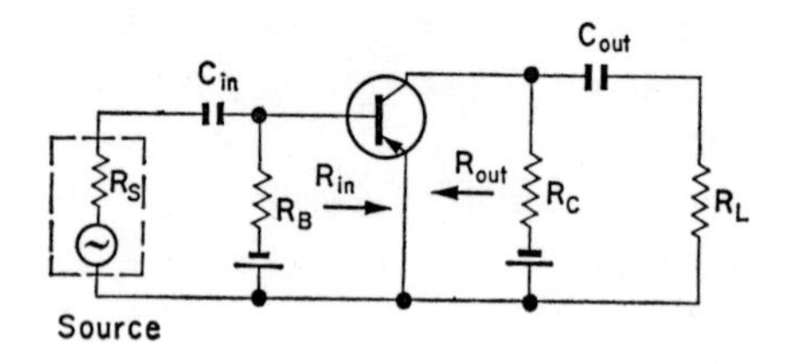

FIG. 6.2.2. Simple common emitter amplifier.

extent on the load resistance R_L, while the output resistances depend on the transistor characteristics and the source resistance R_S.

Typical input characteristics for a low power, alloy junction transistor are shown in Figs. 6.2.3 and 6.2.4. Comparison of these show that for a given change of input voltage, in the common base configuration (Fig. 6.2.3) the current change is of the order of milliamperes while for the common emitter configuration (Fig. 6.2.4) the current change is measured in microamps. Thus the slope resistance is much higher for the latter configuration. The input characteristics are modified

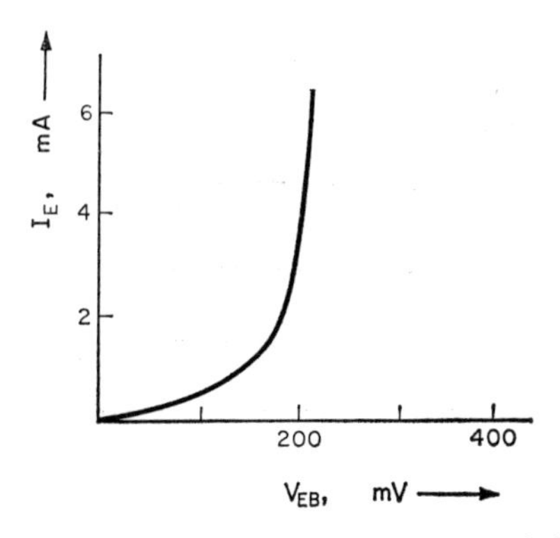

FIG. 6.2.3. Common base input characteristic.

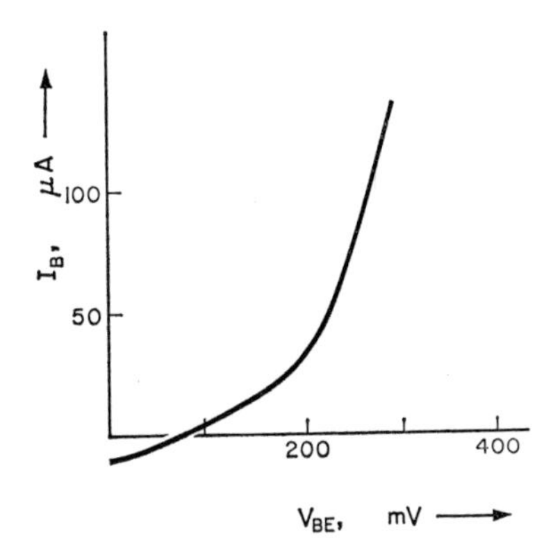

FIG. 6.2.4. Common emitter input characteristic.

slightly by the collector voltage, and hence the collector load resistance. Typical values of input resistance are 50 Ω and 1 kΩ for the common base and common emitter circuits respectively.

The output characteristics for common base and common emitter circuits are shown in Figs. 6.2.5 and 6.2.6 respectively. These indicate the difference in output resistances of the two circuits, the common base characteristics having a very low

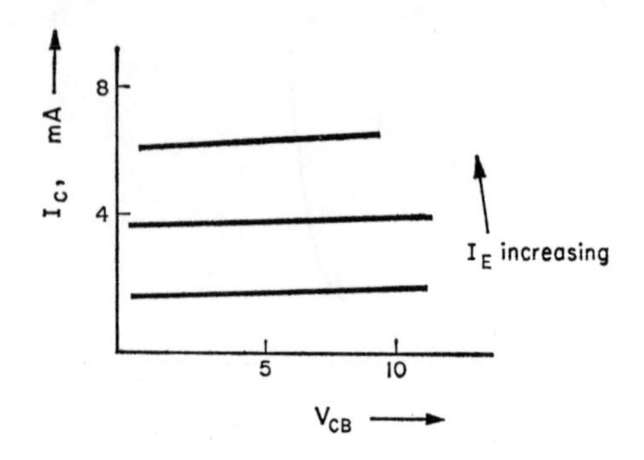

FIG. 6.2.5. Common base output characteristics.

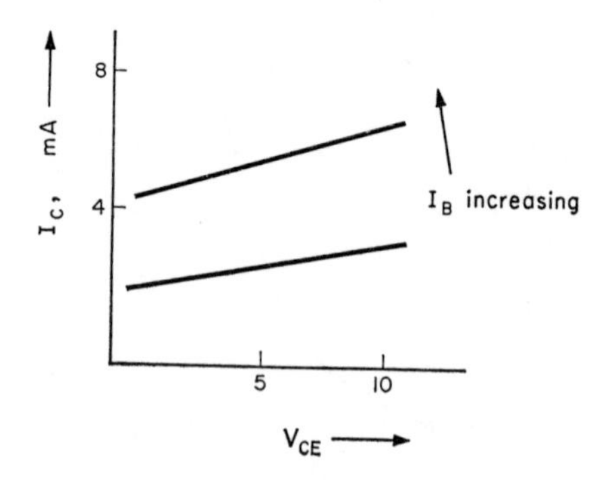

FIG. 6.2.6. Common emitter output characteristics.

slope while the common emitter characteristics are steeper. Typical values of output resistances are 1 MΩ and 20 kΩ respectively.

(b) The cut-off frequencies, i.e. the frequencies where the gain of an amplifier is reduced by a specified amount, occur at frequencies above and below that at which maximum gain is obtained. Low frequency cut-off is not due to the transistor, which can operate at frequencies down to zero (d.c.), but to the external circuit. The reactances of the coupling capacitors C_{in} and C_{out} (Figs. 6.2.1 and 6.2.2) increase as the frequency is reduced decreasing the overall gain. The effect here is similar for both common base and common emitter circuits.

High frequency cut-off is due to the circuit capacitances (particularly the collector capacitance of the transistor itself) and transit time effects within the transistor.

For the common base circuit this occurs when the transit time of the current carriers across the base becomes comparable with the period of the wave being amplified. In the common emitter circuit the base current controls the movement of the current carriers constituting the emitter current and the effect of transit time is magnified. A transistor, whose current gain falls by 3 dB at 100 kc/s in the common base configuration, may have the same reduction at 2 kc/s in the common emitter configuration.

(c) The suitability for being cascaded is mainly the question of matching the output impedance of one stage to the input impedance of the next stage. It is not always possible, or even desirable, to use matching transformers and the high output and low input impedances of the common base circuit render it unsuitable for cascade connection. A.6.14 gives further details of cascade connection of transistor stages.

The term "alpha cut-off frequency" may be defined as that frequency where the current gain falls to $1/\sqrt{2}$ of the low frequency value. It is discussed in several of the questions in Chapters 3 and 4.

Q.6.3

Draw a circuit diagram of a two-stage RC coupled audio-frequency amplifier using transistors. Show typical values for the components and suitable bias and stabilization arrangements.

What factors influence the frequency response of the amplifier you have described?

[C. and G. 2, 1962]

A.6.3

A typical two-stage RC coupled a.f. amplifier is shown in Fig. 6.3.1. Such an amplifier is discussed in A.6.1, the second stage being a replica of Fig. 6.1.1. The first stage is slightly

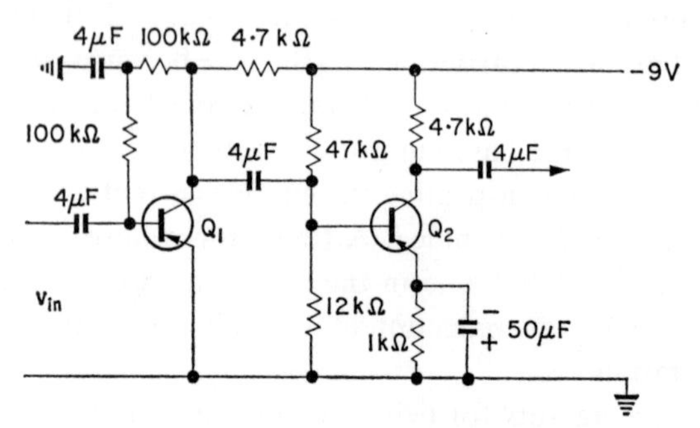

FIG. 6.3.1. Two-stage RC coupled amplifier.

different, the base being fed by a single resistor, and feedback resistor stabilization is employed.

The frequency response of amplifiers is covered in several other questions (e.g. A.4.1 and A.4.2) and will not be covered here.

Q.6.4

Discuss the use of transistors at (a) audio frequencies, and (b) radio frequencies. What is meant by thermal runaway and how is it avoided?

An a.f. RC coupled, grounded emitter transistor amplifier stage has an input impedance of 1000 Ω and a supply of −6 V.

The base voltage is stabilized at $-1\cdot5$ V by a potentiometer consisting of a resistance R and 10,000 Ω connected between base and earth. The collector is fed through a resistance of 1500 Ω and the output taken through a capacitance C. The base current is 30 μA and the current gain of the transistor is 40. Find the value of R, the mid-band current gain and voltage gain of the stage with a load of 1500 Ω. Determine the value of C if the current gain is 3 dB down on mid-band gain at 160 c/s.

[C. and G. 5, 1963]

A.6.4

A discussion on the use of transistors at a.f. and r.f. can obviously have a variety of answers. This solution gives some of the considerations required in each frequency band.

(a) At a.f. the highest frequency used presents little difficulty either in circuit design or the choice of transistor. The main considerations are usually thermal, particularly in the case of power output stages. Thermal instability is due in part to the leakage current that flows in any transistor circuit. In common emitter circuits the leakage current of a voltage amplifier may be an appreciable proportion of the collector current and variation of its value may cause clipping of the wave or under worst conditions thermal runaway.

The latter is due to the current flowing increasing the temperature of the transistor, which in turn increases the leakage current. This cumulative effect can be such as to ruin a transistor.

Suitable stabilization circuits should be used and in the case of high power devices the design of heat sinks is a major consideration.

(b) At r.f. the powers involved are usually smaller and the main considerations are the choice of a transistor with a sufficiently high cut-off frequency and the design of suitable neutralizing networks.

Of the various types of transistor available, and described in detail in A.3.1, the planar epitaxial would give the best results but the cost may be prohibitive.

Neutralizing networks are usually *CR* circuits and feedback energy from the collector to the base circuits. The voltage fed back by the neutralizing network must be in antiphase to that fed back internally through the transistor and its energy transfer sufficient to give a net energy transfer, from output to input

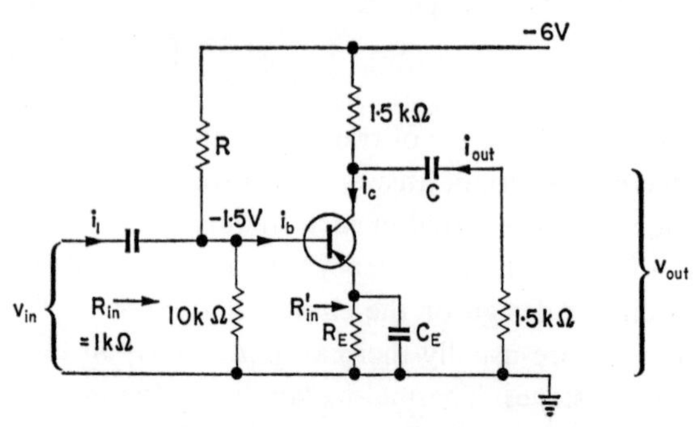

FIG. 6.4.1. Circuit for Q.6.4.

circuits, of zero. The voltage fed back via the neutralizing network is only in antiphase at one particular frequency and care must be taken to avoid oscillation at some other frequency.

The circuit for this particular amplifier is shown in Fig. 6.4.1 emitter resistance and capacity being included to give the complete circuit. The input resistance of 1 $k\Omega$ is assumed to include the shunting effect of the bias circuit.

If a current I flows through the potential divider

$$I = \frac{1{\cdot}5\ \text{V}}{10\ \text{k}\,\Omega} = 0{\cdot}15\ \text{mA}.$$

Therefore

$$R = \frac{4{\cdot}5\ \text{V}}{0{\cdot}15 - 0{\cdot}03\ \text{mA}}$$

$$= \underline{25\ \text{k}\,\Omega.}$$

With the information given we must assume an infinite output impedance. Hence as a current amplifier

$$\text{Overall current gain} = \frac{i_{out}}{i_1}.$$

The shunting effect of the bias network corresponds to a resistance of 25 kΩ in parallel with 10 kΩ, i.e. $\frac{50}{7}$ Ω. Then assuming the input impedance is purely resistive ($= R_{in}$)

$$i_1 \times 1000 = (i_1 - i_b) \times \frac{50}{7} \times 10^3,$$

i.e.

$$i_b = \frac{43}{50}\, i_1.$$

Assuming the reactance of C is negligible the output current splits into two.

Therefore

$$\text{Overall current gain} = \frac{1}{2} \times 40 \times \frac{43}{50}.$$

$$\text{At mid-band} \quad = \underline{17{\cdot}2 \text{ times.}}$$

The mid-band voltage gain

$$= \frac{v_{out}}{v_{in}} = \frac{-i_{out} \times 1500}{i_1 \times 1000}$$

$$= -\ 17{\cdot}2 \times 1{\cdot}5$$

$$= -\ \underline{25{\cdot}8 \text{ times.}}$$

The negative sign signifies phase inversion. Assuming the input and emitter capacitors have a negligible effect, compared with that of the coupling capacitor, the l.f. fall off may be determined from Fig. 6.4.2:

i.e. at mid-band

$$i_{out} = \frac{i_c}{2}.$$

at h.f.

$$i_{out} = \frac{i_c R}{\sqrt{[4R^2 + (1/\omega^2 C^2)]}}$$

$$= \frac{i_c}{2\sqrt{[1 + (1/2\omega CR)^2]}},$$

at the 3 dB part $$i_{\text{out}} = \frac{i_c}{2\sqrt{2}},$$

i.e. $$\frac{1}{2\omega CR} = 1,$$

i.e. $$C = \frac{1}{2\pi \times 160 \times 1{\cdot}5 \times 10^3}$$
$$= \underline{0{\cdot}33\ \mu\text{F}.}$$

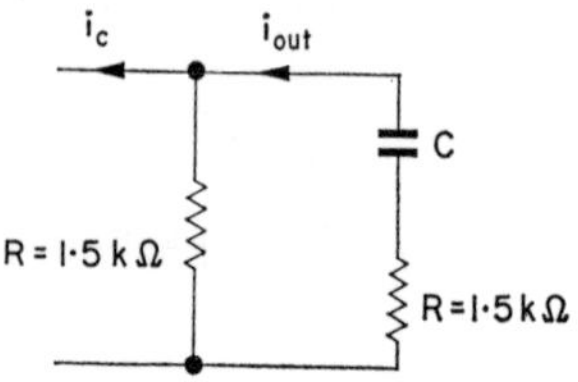

FIG. 6.4.2. Collector output circuit for Q.6.4.

Q.6.5

Explain why grounded emitter amplifiers require some form of d.c. stabilization where grounded base amplifiers are usually unstabilized.

Describe a method of stabilizing the collector current, and derive an expression for the factor of stability of the circuit chosen.

[H.N.C. 3, 1963]

A.6.5

In either the grounded base or the grounded emitter configurations the d.c. collector current may be considered as the sum of two components. One component, caused by the amplifying action is dependent on emitter or base current while the other, having no useful function, is the leakage crurrent which increases considerably with temperature.

The leakage current I_{CO} for a grounded base amplifier is small compared with typical values of collector current. Even if a rise in temperature caused a tenfold increase in leakage current it would still be small compared with the collector current. Hence temperature change causes very little change of the operating point for common base circuits and they are inherently stable.

The worst case for a common emitter circuit is with the base open circuit. The leakage current I_{CO} flowing across the reverse biased collector diode must be supplied from the emitter.

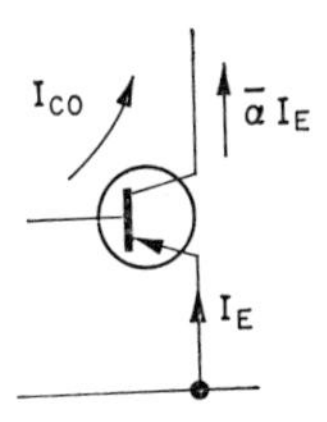

FIG. 6.5.1. Leakage current for common emitter configuration.

Due to the amplifying action of the transistor the total leakage current (see Fig. 6.5.1) is then given by

$$I'_{CO} = I_E = I_{CO} + \bar{\alpha} I_E,$$

where $\bar{\alpha}$ is the d.c. current gain,

or
$$I_E = \frac{I_{CO}}{1-\alpha},$$

$$= I'_{CO}.$$

The value of I'_{CO} may be fifty times that of I_{CO} and a tenfold increase in its value may cause a shift in the operating point that could give rise to distortion, or under worse conditions thermal runaway.

In practice the leakage current for a common emitter circuit lies somewhere between I_{CO} and I'_{CO} depending on the value of the resistance in the base circuit. The need for a means of stabilizing the mean collector current in a common emitter circuit then becomes obvious. The stabilization circuit also allows for variation of the d.c. supply voltage and the current gain of the transistor.

The simplest method of a d.c. stabilization is that using the feedback resistor as shown in Fig. 6.5.2. The operation of the circuit depends on d.c. feedback via the resistor R_B. Any increase in the collector current I_C will reduce the collector voltage. This in turn will cause a reduction in the base emitter

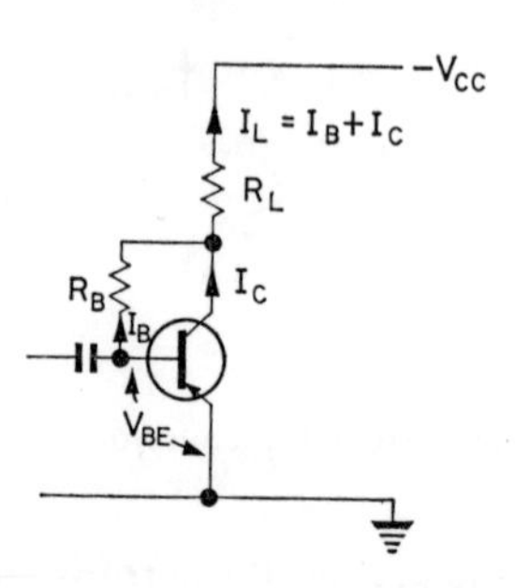

FIG. 6.5.2. Simple d.c. stabilization circuit.

voltage and a decrease in the base current I_B. A fall in I_B leads to a reduction in I_C and the original change in I_C is reduced.

Quantitatively if we ignore the low voltage V_{BE} the following d.c. relations are valid:

$$V_{CC} = (I_B + I_C)R_L + I_B R_B,$$

$$I_C = \bar{\beta} I_B + I'_{CO},$$

where $\bar{\beta}$ is the d.c. current gain.

Eliminating I_B, $$V_{CC} = \left(\frac{I_C - I'_{CO}}{\bar{\beta}}\right)(R_L + R_B) + I_C R_L,$$

i.e. $$I_C\left(1 + \frac{\bar{\beta} R_L}{R_L + R_B}\right) = \frac{\bar{\beta} V_{CC}}{R_L + R_B} + I'_{CO}.$$

Differentiating partially with respect to I'_{CO},

$$\frac{\partial I_C}{\partial I'_{CO}} = \frac{1}{1+[\bar{\beta}R_L/(R_L+R_B)]}.$$

Since ∂I_C is the change in collector current with stabilization and $\partial I'_{CO}$ is the change in collector current without stabilization, this expression gives the factor of stability for the circuit K.

Q.6.6

Compare methods of attaining the correct operating point in valve and transistor circuits. Mention briefly the importance

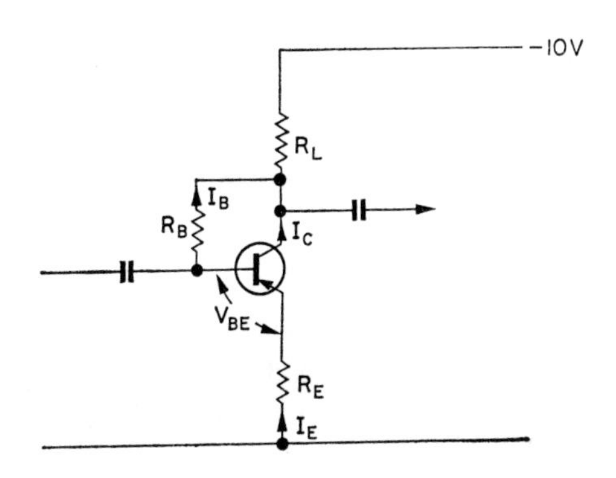

FIG. 6.6.1. Circuit for Q.6.6.

of bias stabilization in transistor circuits. Determine the d.c. operating point for the circuit shown in Fig. 6.6.1.

Comment on the circuit and note all assumptions made.

$$I_{CO} = 10\ \mu\text{A}, \quad \alpha_0 = 0{\cdot}98, \quad R_L = 3{\cdot}3\ \text{k}\Omega,$$

$$R_B = 220\ \text{k}\Omega, \quad R_E = 470\ \Omega$$

[I.E.E., 1963]

A.6.6

In valve circuits the operating point (given by values of anode current and voltage) is determined for a particular supply voltage by the d.c. potential of the grid relative to the cathode. This potential difference is conveniently produced across a cathode resistor as shown in Fig. 6.6.2. It should be noted that the bias condition is one of voltage.

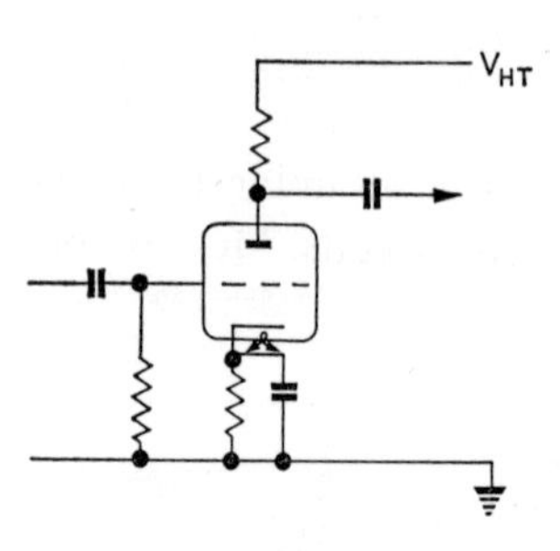

FIG. 6.6.2. Bias arrangements for a valve circuit.

For transistor circuits the operating point (given by values of collector current and voltage) is determined for a given collector supply voltage by the steady value of base or emitter current, depending on the circuit configuration. The bias condition is one of current, and typical circuits are given in this chapter.

The importance of bias stabilization is discussed in the previous solution (A.6.5).

For the circuit given in Fig. 6.6.1:

$$\bar{\beta} = \frac{\alpha_0}{1-\alpha_0} = 49$$ (assuming the d.c. current gain equals the a.c. current gain),

$$I'_{CO} = I_{CO}(1+\bar{\beta}) = 500\ \mu\text{A} = 0{\cdot}5\ \text{mA},$$

$$I_C = \bar{\beta} I_B + I'_{CO} = 49 I_B + 0{\cdot}5. \qquad (1)$$

If we ignore the low voltage V_{BE},

$$\begin{aligned} 10 &= I_E \times 0{\cdot}47 + I_B \times 220 + (I_B + I_C)\, 3{\cdot}3 \\ &= 223{\cdot}8\, I_B + 3{\cdot}77 I_C \end{aligned} \quad (2)$$

since $I_E = I_B + I_C$.

Eliminating I_C between (1) and (2),

$$10 = 223{\cdot}8\, I_B + 3{\cdot}77\,(49\, I_B + 0{\cdot}5),$$

i.e. $\underline{I_B = 19{\cdot}7\ \mu\text{A}.}$

Therefore $\underline{I_C = 1{\cdot}47\ \text{mA}}$

and $V_C = 10 - 1{\cdot}47 \times 3{\cdot}3 = \underline{5{\cdot}15\ \text{V}}$

$V_E = 0{\cdot}47 \times 1{\cdot}49 = \underline{0{\cdot}7\ \text{V}.}$

The circuit shown will give a low value of gain due to feedback via the base resistor R_B and the emitter resistor R_E. The latter is usually decoupled by a high value capacitor and R_B split with its mid-point earthed through a smaller capacitor.

Q.6.7

Show from a consideration of the equations

$$I_C = \alpha I_E + I_{CO}, \qquad I_E = I_B + I_C,$$

that the stability criterion $S = \partial I_C / \partial I_{CO}$ for a transistor is given by

$$S = \frac{1 + (R_E/R_B)}{1 - \alpha + (R_E/R_B)},$$

where R_B and R_E are the total resistance of the base and emitter circuits respectively.

A common emitter amplifier uses a transistor having $\beta = 48$. The load resistance is 5 kΩ and the maximum tolerable emitter lead resistance is 1 kΩ and the collector emitter voltage must not vary by more than 1 V. Due to temperature changes, the value of I_{CO} is liable to change from 5μA to 40 μA.

What restriction does this place on the choice of the base circuit resistance? [B.Sc. 3, 1960]

A.6.7

The particular bias circuit is not given in this question and the circuit assumed is given in Fig. 6.7.1. By the use of Thevenin's theorem any bias circuit may be reduced to this form.

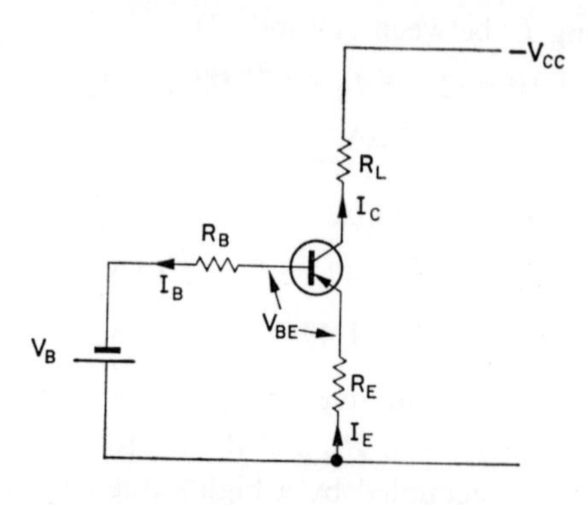

FIG. 6.7.1. Circuit for Q.6.7.

If we ignore the low voltage drop V_{BE},

$$V_B = I_E R_E + I_B R_B$$

$$= -I_C R_B + I_E(R_E + R_B), \quad \text{since } I_E = I_B + I_C.$$

Also

$$I_C = \alpha I_E + I_{CO}.$$

Therefore eliminating I_E,

$$V_B = -I_C R_B + \left(\frac{I_C - I_{CO}}{\alpha}\right)(R_E + R_B)$$

or

$$I_C\left(1 - \frac{\alpha R_B}{R_E + R_B}\right) = \frac{\alpha V_B}{R_E + R_B} + I_{CO}.$$

Therefore

$$S = \frac{\partial I_C}{\partial I_{CO}} = \frac{1}{1 - [\alpha R_B/(R_E + R_B)]} = \frac{R_E + R_B}{(1-\alpha)R_B + R_E}$$

$$= \frac{1 + R_E/R_B}{1 - \alpha + R_E/R_B}.$$

Also $$V_{CE} = -V_{CC}+I_C R_L+I_E R_E$$

$$= -V_{CC}+I_C R_L+\left(\frac{I_C-I_{CO}}{\alpha}\right) R_E$$

$$= -V_{CC}+\left\{\frac{[\alpha V_B/(R_E+R_B)]+I_{CO}}{1-[\alpha R_B/(R_E+R_B)]}\right\}\left(R_L+\frac{R_E}{\alpha}\right)-\frac{I_{CO}R_E}{\alpha}.$$

Therefore $$\frac{\partial V_{CE}}{\partial I_{CO}} = \frac{(R_L+(R_E/\alpha)]}{1-[\alpha R_B/(R_E+R_B)]}-\frac{R_E}{\alpha}$$

$$= \frac{R_L+[R_E R_B/(R_E+R_B)]}{1-(\alpha R_B/(R_E+R_B)]},$$

i.e. $$\frac{\partial V_{CE}}{\partial I_{CO}} = \frac{R_L(R_E+R_B)+R_E R_B}{R_E+R_B(1-\alpha)}$$

$$= \frac{R_E+R_L[1+(R_E/R_B)]}{1-\alpha+R_E/R_B}$$

In the example $$\frac{\partial V_{CE}}{\partial I_{CO}} \ngtr \frac{1}{3{\cdot}5\times 10^{-6}},$$

i.e. $$\frac{1000+5000[1+(R_E/R_B)]}{1/49+(R_E/R_B)} \ngtr \frac{1}{35\times 10^{-6}},$$

or $$0{\cdot}21+0{\cdot}175\frac{R_E}{R_B} \ngtr 0{\cdot}0204+\frac{R_E}{R_B},$$

i.e. $$0{\cdot}19 \ngtr 0{\cdot}825\frac{R_E}{R_B},$$

i.e. $$R_B \ngtr 4{\cdot}34\ R_E.$$

Hence under the conditions given R_B must not be greater than 4.34 kΩ. This would probably mean that a potential divider would supply the base bias.

Q.6.8

Explain why the collector current in a grounded base junction transistor can be expressed as $I_{CO}+\bar{\alpha}I_E$ and derive the corresponding expression for a grounded emitter transistor.

In the circuit shown (Fig. 6.8.1) the current I is provided by a low impedance source, when I is zero it is found that $I_1 = 2{\cdot}6$ mA. Calculate the value of the current I_{CO} if the current gain is 0·96. What is the value of the current I if the input current is 0·1 mA? Assume in both calculations that the voltage between the base and the emitter is negligible.

[B.Sc. 3, 1959]

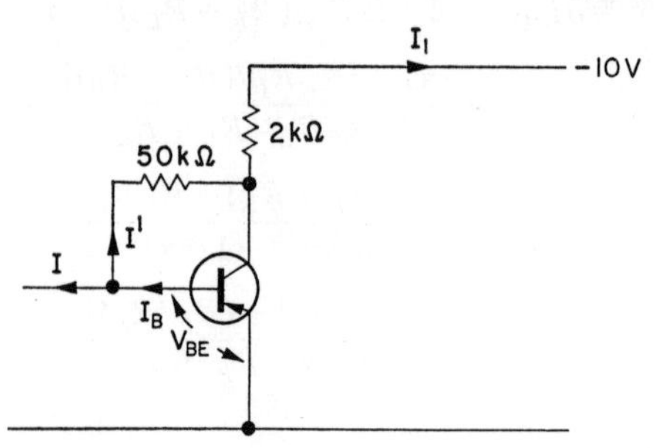

FIG. 6.8.1. Circuit for Q.6.8.

A.6.8

As explained in A.6.5 the total collector current is made up of two components, one due to the amplifying action of a transistor the other due to the leakage current.

Hence the total current flowing out of the collector in the grounded base configuration is given by

$$I_C = \bar{\alpha} I_E + I_{CO},$$

where $\bar{\alpha}$ is the fraction of the emitter current appearing at the collector junction and I_{CO} is the leakage current flowing through the reverse biased collector diode.

For the common emitter circuit the base current is the controlling factor.

Hence eliminating I_E between the equations

$$I_C = \bar{\alpha} I_E + I_{CO} \quad \text{and} \quad I_E = I_B + I_C$$

gives $$I_C = \bar{\alpha}(I_B+I_C)+I_{CO}$$

or $$I_C = \frac{\bar{\alpha}}{1-\bar{\alpha}} I_B+I'_{CO}$$

$$= \bar{\beta}I_B+I'_{CO}$$

where $$\bar{\beta} = \frac{\bar{\alpha}}{1-\bar{\alpha}} \quad \text{and} \quad I'_{CO} = \frac{I_{CO}}{1-\bar{\alpha}}.$$

In the example

$$\bar{\alpha} = 0{\cdot}96,$$

$$\bar{\beta} = \frac{0{\cdot}96}{1-0{\cdot}96} = 24.$$

If we ignore the low voltage V_{BE},

$$10 = I_1\, 2+50I' \quad (I' \text{ and } I_B \text{ in mA}). \tag{1}$$

But $$I_C = (I_1-I') = 24\, I_B+I'_{CO}$$

$$= 24\,(I+I')+I'_{CO}. \tag{2}$$

Hence from (2)

$$I' = \frac{I_1-24I-I'_{CO}}{25}$$

and from (1) $$10 = I_1 2+50\,\frac{(I_1-24I-I_{CO})}{25}$$

$$= 4I_1-4{\cdot}8I-2I_{CO}$$

when $$I=0,\; I_1 = 2{\cdot}6.$$

Therefore $$10 = 10{\cdot}4-2I_{CO},$$

i.e. $$I'_{CO} = 0{\cdot}2 \text{ mA}.$$

Therefore $$I_{CO} = 0{\cdot}2(1-0{\cdot}96)$$

$$= \underline{0{\cdot}008 \text{ mA.}}$$

When $$I = 0{\cdot}1 \text{ mA},$$

$$10 = 4I_1-4{\cdot}8-0{\cdot}4,$$

i.e. $$I_1 = \frac{15{\cdot}2}{4}$$

$$= \underline{3{\cdot}8 \text{ mA.}}$$

Q.6.9

Measurements made on a transistor produced the following results:

Common emitter connection

V_{CE}	(V)	−2	−2	−2	−2	−2	−8	−8	−8	−8	−8
I_C	(mA)	0·1	1·0	2·0	3·0	4·0	0·3	1·4	2·6	3·9	5·2
I_B	(μA)	0	20	40	60	80	0	20	40	60	80

Common base connection

V_C	(V)	−2	−2	−2	−2	−2	−8	−8	−8	−8	−8
I_C	(mA)	1·0	3·0	4·9	6·8	8·8	1	3	4·9	6·8	8·8
I_E	(mA)	1	3	5	7	9	1	3	5	7	9

Draw and compare the complete output characteristic curves for each method of connection over the range of collector voltage from 0 to −10 V.

On the common emitter curves, construct load lines for 2 kΩ and 500 Ω using a 6 V supply.

If the input were an alternating current of 60 μA peak-to-peak, choose a suitable steady base current and calculate the current amplification for each value of load.

[H.N.D. 2, 1962]

A.6.9

The output characteristics for the two configurations are shown in Figs. 6.9.1 and 6.9.2. The main difference between the two sets of curves is the difference in slope resistances. To the accuracy of measurement the common base characteristics have a slope resistance of infinity, while the common emitter characteristics have a slope resistance of about 6·7 kΩ. A minor difference, not shown by the results, is that for the common base configuration the collector must be taken positive to reduce the current to zero.

A load line may be defined as the variation of collector voltage V_C with collector current I_C and is therefore dependent

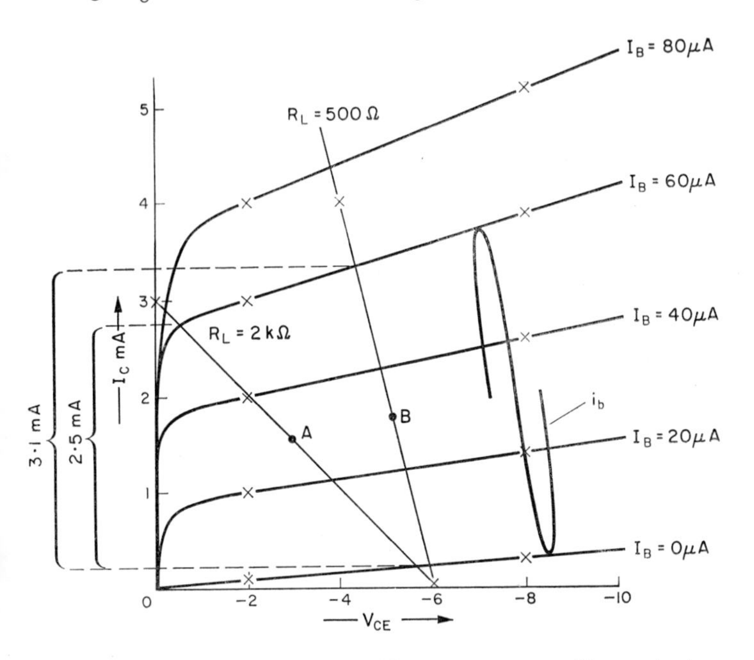

FIG. 6.9.1. Load lines for Q.6.9 on common emitter output characteristics.

on the voltage drop in the collector load resistance. If the d.c. load resistance between the collector supply voltage $-V_{CC}$ and the collector is R_L, the d.c. load line is given by

$$V_C = -V_{CC}+I_C \times R_L.$$

If the a.c. load impedance is the same as the d.c. load resistance, the above equation also represents the a.c. load line.

The above holds for *pnp* transistors, but for *npn* transistors the sign of V_{CC} and $I_C R_L$ are reversed.

Hence the load lines for the resistances given are:

$$V_C = -6+2I_C \quad (I_C \text{ in mA})$$

and

$$V_C = -6+0{\cdot}5\, I_C \quad (I_C \text{ in mA}).$$

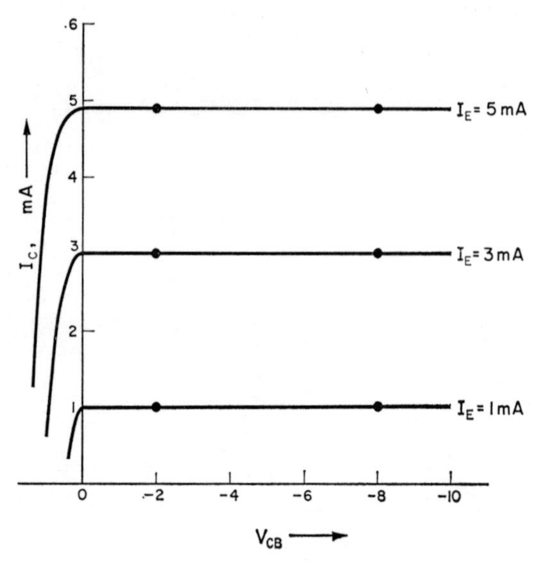

FIG. 6.9.2. Common base characteristics for Q.6.9.

These may be plotted on the characteristics shown in Fig. 6.9.1, i.e. for the 2 kΩ load line.

If $I_C = 0$, $V_C = -6$ V; if $V_C = 0$, $I_C = -3$ V.

For the 500 Ω load line:

If $I_C = 0$, $V_C = -6$ V; if $I_C = 4$ mA, $V_C = -4$ V.

For the 2 kΩ load line a 60 μA peak-to-peak signal is just on the limit of distortionless reproduction. The operating point would be at A (i.e. 30 μA base current) and corresponds to a collector voltage of $-2{\cdot}7$ V and a collector current of $1{\cdot}55$ mA.

The variation in collector current for a base current swing of 60 μA is 2·5 mA.

Hence the current gain is $\dfrac{2\cdot5 \text{ mA}}{60 \ \mu\text{A}} = 42$ times.

For the 500 Ω load a suitable operating point would correspond to B. The standing base current is agin 30 μA while the collector voltage and current are $-5\cdot15$ V and 1·75 mA respectively.

The variation in collector current for a swing of 60 μA in the base current for the 500 Ω load is then 3·1 mA and the current gain is

$$\frac{3\cdot1 \text{ mA}}{60 \ \mu\text{A}} = \underline{52} \text{ times.}$$

Q.6.10

The collector current–collector voltage characteristics of a junction transistor in common emitter connection, at different values of base current, are given in Table 6.1.

TABLE 6.1

$-V_C$ (V)	$-I_C$ mA		
	$I_B = -20\ \mu$A	$I_B = -40\ \mu$A	$I_B = -60\ \mu$A
0·25	0·30	0·90	1·20
0·5	0·35	1·30	2·20
1	0·40	1·40	2·40
3	0·50	1·50	2·50
6	0·65	1·65	2·65

The transistor is used as a common emitter amplifier with a battery supply of 6 V and a collector load resistance of 2 kΩ. The quiescent operating point is $V_C = -3$ V, $I_C = -1\cdot5$ mA. Plot the above curves and draw the load line. If an input signal of 50 mV peak causes the base current to change by 20 μA, determine the current gain and the voltage gain.

[H.N.C., 1962]

A.6.10

This question is very similar to the previous one and the load line, represented by $V_C = -6+2I_C$, is shown superimposed on the output characteristics in Fig. 6.10.1.

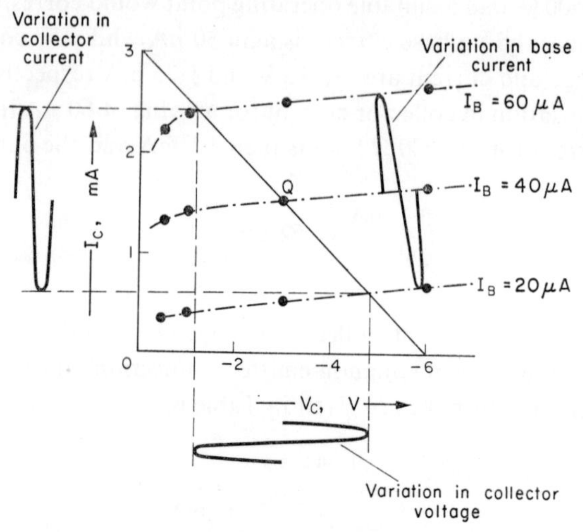

FIG. 6.10.1. Load line for Q.6.10.

From the figure:

$$\text{Current gain} = \frac{\text{variation in output current}}{\text{variation in input current}}$$

$$= \frac{2{\cdot}4-0{\cdot}6 \text{ mA}}{40\ \mu\text{A}}$$

$$= \underline{45 \text{ times.}}$$

$$\text{Voltage gain} = \frac{\text{variation in output voltage}}{\text{variation in input voltage}}$$

$$= \frac{4{\cdot}8-1{\cdot}2 \text{ V}}{100 \text{ mV}}$$

$$= \underline{36 \text{ times.}}$$

Q.6.11

Characteristics of a junction transistor are given in Table 6.2.

TABLE 6.2

Collector volts (V_{CE})	Collector current (I_C mA)		
	$I_B = 0$	$I_B = 40\ \mu A$	$I_B = 80\ \mu A$
1·0	0·20	1·90	3·7
4·0	0·30	2·05	4·0
7·0	0·40	2·20	4·3

The transistor is connected in a common-emitter stage with a collector load of 1500 Ω, a supply voltage of 6 V and a d.c. bias of 40 μA.

Plot the characteristics and draw the appropriate load line. Calculate the power dissipated in the transistor. What will be the total voltage swing at the collector for an a.c. input signal current of 40 μA peak in the base?

[C. and G. 2, 1961]

A.6.11

The calculation of the total voltage swing in this question is very similar to those of the previous two questions. It is left to the reader to show that a voltage swing of 5 V is obtained.

The power dissipated in the transistor may be determined by finishing the d.c. power available at the collector and subtracting the a.c. power developed at the collector (see A.7.2). The reader is left to show that 4 mW is dissipated at the collector.

Q.6.12

Measurements made on a small *pnp* transistor gave the following results when connected as a common emitter circuit.

V_{CE}	−2 V					−8 V				
I_C mA	0·1	1	2	3	4	0·3	1·4	2·6	3·9	5·4
I_B μA	0	20	40	60	80	0	20	40	60	80

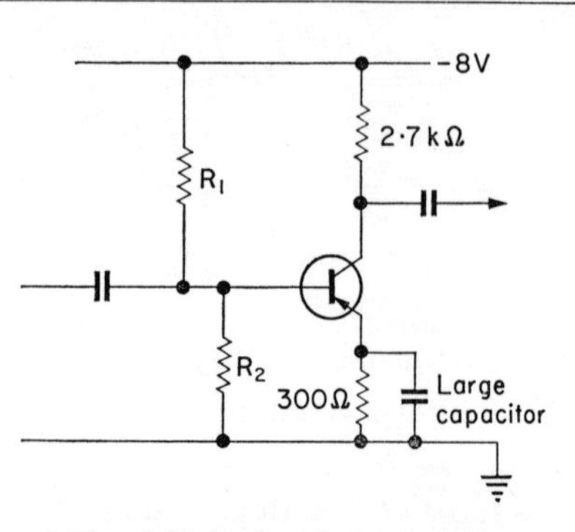

FIG. 6.12.1. Circuit for Q.6.12.

It is connected in the circuit shown in Fig. 6.12.1 and biased such that the d.c. base input current is 20 μA. If the input current is 20 μA peak, determine the output voltage when working into a high impedance load.

Determine also the overall current gain if it is connected to another stage of input resistance 900 Ω by a capacitor of very low reactance.

[H.N.C. 3, 1964]

A.6.12

The output characteristics are shown in Fig. 6.12.2 and superimposed on them are the three load lines required for this question.

To determine the d.c. operating point, the total resistance between the supply line and earth must be considered:

i.e. $$V_{CE} = -8+(2{\cdot}7+0{\cdot}3)I_C \quad (I_C \text{ in mA})$$

or $$V_{CE} = -8+3I_C,$$

i.e. the d.c. load line AB passes through the points $I_C = 0$, $V_{CE} = -8$ V and $V_{CE} = 0$, $I_C = 2{\cdot}67$ mA, and where this cuts the 20 μA characteristic gives the operating point Q (i.e. $V_{CE} = -4{\cdot}5$ V, $I_C = 1{\cdot}17$ mA).

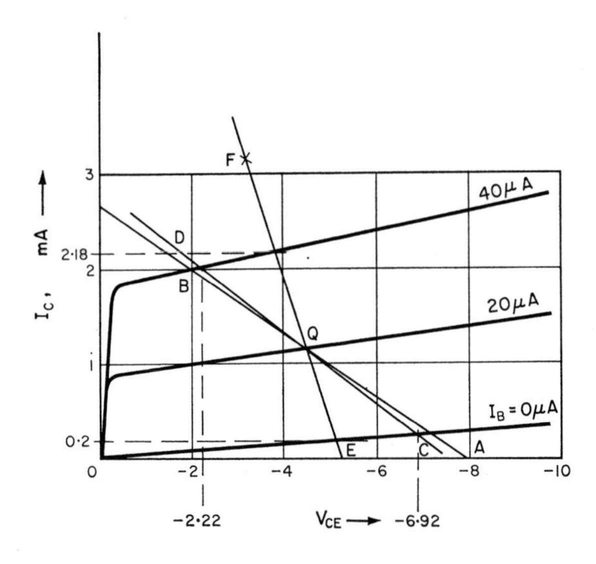

FIG. 6.12.2. Output characteristics and load lines for Q.6.12.

With a high impedance load the a.c. impedance between collector and emitter is 2·7 kΩ. This means a change in current of 1 mA would give a rise to a voltage change of 2·7 V. Hence the a.c. load line CD may be drawn corresponding to this slope and passing through the point Q. From the intersections of this load line and the characteristics it can be seen that a swing of ± 20 μA in the base current gives rise to a swing of output voltage of $-6{\cdot}92$ V to $-2{\cdot}2$ V, i.e. 4·7 V *pp* or

$$\frac{4{\cdot}7}{2\sqrt{2}} = \underline{1{\cdot}66 \text{ V r.m.s.}}$$

With a 900 Ω load connected via a low reactance capacitor the effective load impendance is 900 Ω in parallel with 2·7 Ω, i.e. 675 Ω. A second a.c. load line EF may be drawn having a slope of 675 Ω and passing through the point Q. The intersections of this load line and the characteristics gives the swing in collector current, i.e. a 40 μA swing in base current gives rise to a collector current swing of $2{\cdot}18-0{\cdot}2 = 1{\cdot}98$ mA.

The alternating current at the collector divides into two, 75% of it flowing through the 900 Ω resistor and 25% through the 2·7 kΩ resistor.

Hence the overall current gain is

$$\frac{3}{4}\times\frac{1{\cdot}98 \text{ mA}}{40\ \mu\text{A}}$$

$$= \underline{37 \text{ times.}}$$

Q.6.13

The circuit shown, Fig. 6.13.1 shows a common-emitter transistor amplifier for Class A operation.

Determine values for R_L, R_1, and C_1, if the mean collector current is 2 mA and a loss of 3 dB can be tolerated at 50 c/s. Assume α to be 0·98 and the resistance of input to be 1 kΩ.

[B.Sc. 3, 1960]

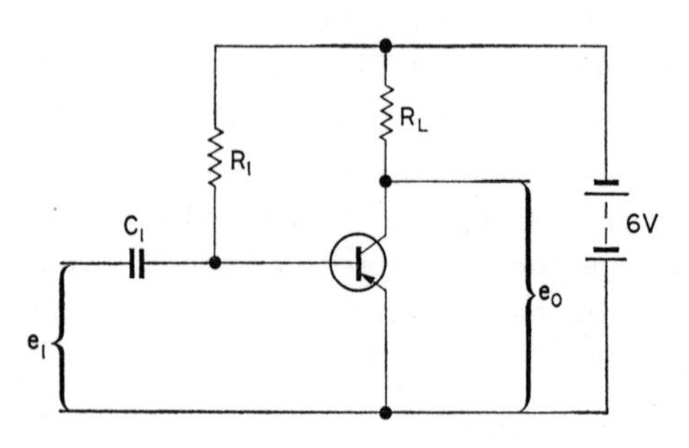

FIG. 6.13.1. Circuit for Q.6.13.

A.6.13

Due to the limited information given, several assumptions have to be made in the solution of this problem.

Assuming that the current gain is the same for a.c. as d.c.,

$$\bar{\beta} = 1-\frac{0\cdot98}{0\cdot98}$$

$$= 49.$$

Ignoring the leakage current,

$$I_B = \frac{2}{49} = 0\cdot0408 \text{ mA}$$

$$= \underline{40\cdot8\ \mu\text{A}.}$$

Hence assuming that the d.c. voltage between base and emitter is very small,

$$R_1 = \frac{6 \text{ V}}{40\cdot8\ \mu\text{A}} = 147 \text{ k}\Omega.$$

Taking the nearest standard value $R_1 = 150$ kΩ. This gives an error in the right direction to allow for the leakage current.

Assuming a minimum collector emitter voltage of about 1 V and a collector current swing of 0 to 4 mA.

$$\text{Collector load resistance } R_L = \frac{5 \text{ V}}{4 \text{ mA}}$$

$$= 1\cdot25 \text{ k}\Omega.$$

The nearest standard value is 1·2 kΩ. At 50 c/s the fall off would be due to the coupling capacitor C_1 and at the 3 db point

$$\frac{R_{\text{in}}}{\sqrt{(X_C^2+R_{\text{in}}^2)}} = \frac{1}{\sqrt{2}}$$

or

$$X_C = R_{\text{in}}.$$

Therefore
$$C = \frac{1}{2\pi f R_{in}}$$
$$= \frac{1}{2\pi 50 \times 10^3}$$
$$= 3{\cdot}18 \ \mu F.$$

In practice a 4 μF capacitor would be suitable, care being required to ensure the d.c. potential correctly polarizes the capacitor.

Q.6.14

Write brief notes on methods of cascading transistor amplifier stages.

An instrument amplifier has five identical stages in cascade. If the overall gain at 100 kc/s is 0·5 dB down on the mid-band gain, calculate the upper 3 dB cut-off frequency for an individual stage. Derive all formulae used and mention all assumptions.

[I.E.E., Nov. 1963]

A.6.14

The main consideration in cascading transistor amplifier stages is correct matching of input and output impedances. In practice multistage amplifiers use the common emitter configuration, since the wide variation between input and output impedances render common base and common collector circuits unsuitable. Even with common emitter circuits coupling transformers are often desirable to help correct matching of input and output. The higher gain possible with common emitter stages is a further point in its favour.

The question gives no information about the amplifier itself but it is reasonable to assume the gain of each stage may be represented by

$$m = \frac{A}{1+jKf} \qquad \text{or} \qquad |m| = \frac{A}{\sqrt{(1+K^2f^2)}},$$

where A is the l.f. gain, K is a constant depending on the amplifier, f is the frequency of operation.

The overall gain is then

$$m^5 = \frac{A^5}{(1+jKf)^5} \qquad \text{or} \qquad |m| = \frac{A^5}{(1+K^2f^2)^{5/2}};$$

$$\text{at 100 kc/s } 0{\cdot}5 = 20 \log_{10} \frac{A^5}{A^5/(1+K^2\times 10^{10})^{5/2}}$$

$$= 50 \log_{10} (1+K^2\times 10^{10}),$$

$$\text{i.e. } 1+K^2 10^{10} = \text{antilog } 0{\cdot}01$$

$$= 1{\cdot}0233.$$

$$\text{Therefore } K^2 = 0{\cdot}0233\times 10^{-10}$$

$$\text{or } K = 1{\cdot}53\times 10^{-6}.$$

For a single stage the 3 dB point corresponds to the gain falling to 0·707 of the low frequency value:

$$\text{i.e. } Kf_{CO} = 1$$

$$\text{or } f_{CO} = \frac{1}{1{\cdot}53\times 10^{-6}}$$

$$= \underline{654 \text{ kc/s.}}$$

Q.6.15

Calculate the mid-band current gain of a two-stage common emitter transistor amplifier. Each stage has a 1 kΩ load resistor and a 100 kΩ resistor is connected between the base and the negative supply to determine the operating point.

The constants of the *pnp* transistor are: $r_e = 30\ \Omega$, $r_b = 1\ \text{k}\Omega$, $r_c = 1\ \text{M}\Omega$ and $\alpha = 0{\cdot}98$.

Derive all formulae used and draw a circuit diagram.

[I.E.E., May 1962]

A.6.15

The circuit of such an amplifier is shown in Fig. 6.15.1 and the equivalent circuit in Fig. 6.15.2. In the latter the shunting effect of the 100 kΩ bias resistor has been ignored since this would be very much greater than the input resistance of the transistor.

The loop equations may then be written

$$i_1(10^6+10^3+30)+0{\cdot}98(-i_{in}-i_1)\times 10^6+i_{in}30+i_2\times 10^3 = 0$$

or $$i_1\times 21\times 10^3+i_2\cdot 10^3-0{\cdot}98\times 10^6 i_{in} = 0,$$

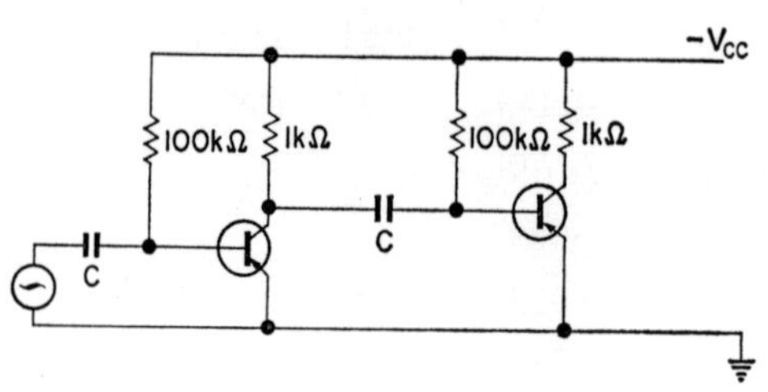

FIG. 6.15.1. Circuit for Q.6.15.

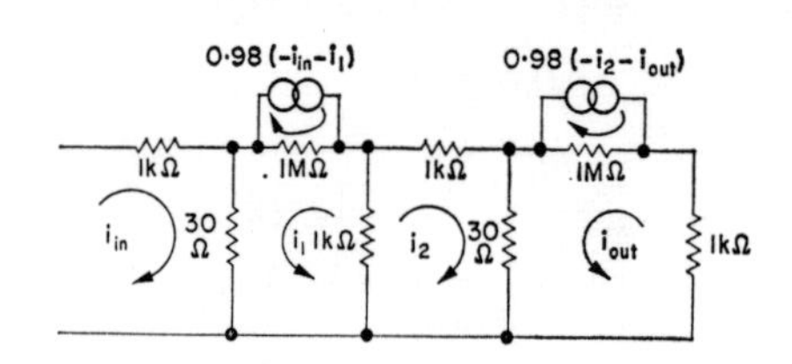

FIG. 6.15.2. Equivalent T circuit for Q.6.15.

i.e. $$21i_1+i_2-980i_{in} = 0 \quad (1)$$

$$i_2(10^3+10^3+30)+i_1\cdot 10^3+i_{out}\cdot 30 = 0$$

or $$2{\cdot}03\times i_2+i_1+0{\cdot}03i_{out} = 0 \quad (2)$$

$$i_{out}(10^6+10^3+30)+0{\cdot}98(-i_{out}-i_2)10^6+i_2 30 = 0$$

or $$i_{out}\cdot 21\times 10^3-0{\cdot}98\times 10^6 i_2 = 0,$$

i.e. $$21i_{out}-980i_2 = 0. \quad (3)$$

From (3) $$i_2 = 0{\cdot}0214\, i_{out}.$$

From (2) $$i_1 = -2{\cdot}03\times 0{\cdot}0214 i_{out}-0{\cdot}03 i_{out}$$
$$= -0{\cdot}0735_{out}.$$

From (1) $$980 i_{\text{in}} = 21(-0{\cdot}0735 i_{\text{out}}) + 0{\cdot}214 i_{\text{out}}$$
$$= 1{\cdot}52 i_{\text{out}}.$$

Therefore current gain $$= \frac{i_{\text{out}}}{i_{\text{in}}}$$
$$= -\frac{980}{1{\cdot}52}$$
$$= \underline{-645 \text{ times.}}$$

The negative sign indicates that the direction of i out must be reversed and shows that an increase in current flowing into the amplifier gives rise to an increase in current flowing out of the amplifier.

Q.6.16

Explain the meaning of the terms *pnp* and *npn* as applied to transistors. What is the essential difference in the circuits appropriate to these two types of transistor?

The characteristics of a transistor in grounded base connection are, for small signals, represented by the following equations:

$$i_c = \alpha i_e - g_c v_c,$$
$$v_e = r_e i_e + \mu v_c.$$

In a particular transistor,

$$\alpha = 0{\cdot}97, \quad g_c = 5\ \mu\text{mho}$$
$$r_e = 20\ \Omega, \quad \mu = 0{\cdot}0005$$

If this transistor is used in grounded base connection with a load resistor of 50 kΩ in the collector circuit, calculate (a) the voltage amplifications, (b) the input resistance.

[C. and G. 5, 1962]

A.6.16

The terms *pnp* and *npn* refer to the type of semiconductor material used for the emitter, base and collector regions of transistors. In *pnp* transistors the emitter region is *p*-type semiconductor material and the majority current carries are holes, while for *npn* the emitter is made of *n*-type material and the current is carried by electrons.

In each type the emitter diode is forward biased and the collector junction reverse biased. Hence the polarities of the d.c. supplies are different for the two types and typical common base amplifiers are shown in Figs. 6.16.1 and 6.16.2.

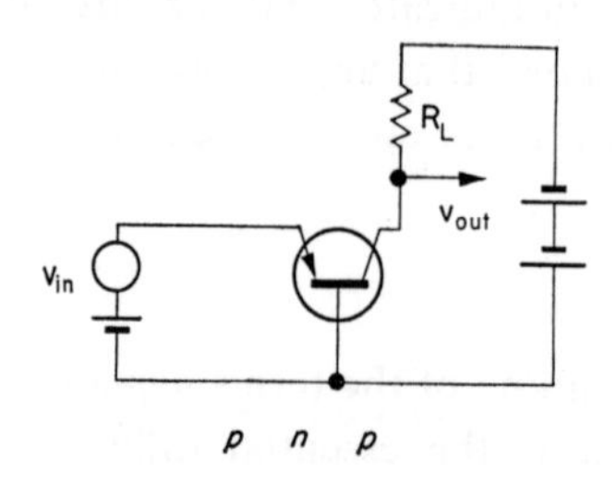

FIG. 6.16.1. Common base amplifier with *pnp* transistor.

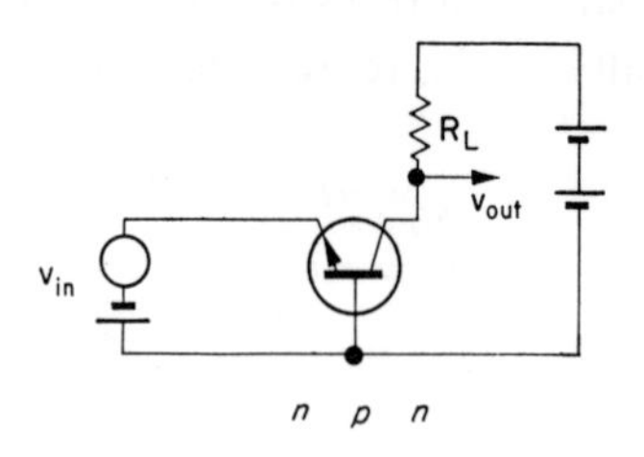

FIG. 6.16.2. Common base amplifier with *npn* transistor.

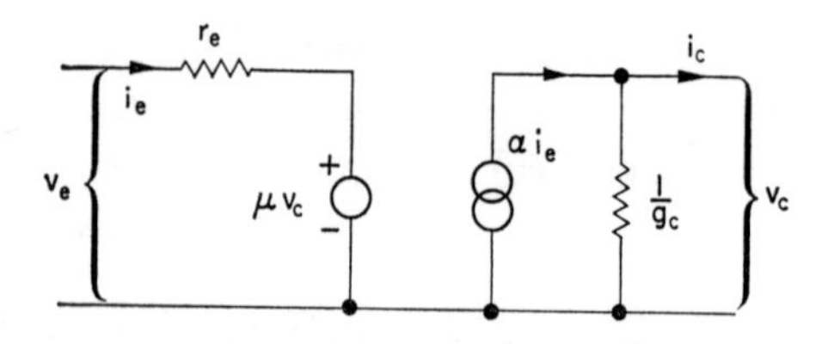

FIG. 6.16.3. Equivalent circuit for Q.6.16.

The equations representing the transistor characteristics show that (a) the collector current (i_c) is a function of the emitter current (i_e) and the collector voltage (v_c); (b) the emitter voltage (v_e) is a function of the emitter current (i_e) and collector voltage (v_c).

Hence the small signal equivalent circuit is as shown in Fig. 6.16.3. The circuit equations are:

$$v_e = 20i_e + 0{\cdot}0005\, v_c, \quad (1)$$

$$i_c = 0{\cdot}97i_e - 5\times 10^{-6}\, v_c, \quad (2)$$

$$v_c = i_c\, 50000. \quad (3)$$

From (2) and (3)

$$\frac{v_c}{50000} = 0{\cdot}97i_e - 5\times 10^{-6} v_c.$$

Therefore $\quad v_c(20\times 10^{-6} + 5\times 10^{-6}) = 0{\cdot}97i_e,$

i.e. $\quad v_c = \dfrac{0{\cdot}97\times 10^6 i_e}{25}.$

Hence from (1) $\quad v_e = 20i_e + 0{\cdot}0005\times \dfrac{0{\cdot}97\times 10^6 i_e}{25}$

$$= 39{\cdot}4i_e.$$

Therefore input resistance $= \dfrac{v_e}{i_e} = \underline{39{\cdot}4\ \Omega}$

and voltage amplification

$$\frac{v_c}{v_e} = \frac{0{\cdot}97\times(10^6/25)\times i_e}{39{\cdot}4 i_e}$$

$$= \underline{985 \text{ times.}}$$

Q.6.17

Prove the relations given in Table 6.3, stating any assumptions made.

TABLE 6.3

	Common base	Common emitter	Common collector
Input impedance	$r_e+r_b(1-\alpha_0)$	$r_b+\dfrac{r_e}{1-\alpha_0}$	$r_b+\dfrac{R_L}{(R_L/r_c)+(1-\alpha_0)}$
Output impedance	$r_c-\dfrac{\alpha_0 r_c r_b}{r_e+R_s+r_b}$	$r_c(1-\alpha_0)+\dfrac{\alpha_0 r_c r_e}{R_S+r_b+r_e}$	$r_e+(R_S r_b)(1-\alpha_0)$
Current gain	$-\dfrac{\alpha_0}{1+(R_L/r_c)}$	$\dfrac{\alpha_0}{(R_L/r_c)+(1-\alpha_0)}$	$-\dfrac{1}{(1-\alpha_0)+(R_L/r_c)}$
Voltage gain	$\dfrac{\alpha_0 R_L}{r_e+r_b(1-\alpha_0)}$	$-\dfrac{\alpha_0 R_L}{r_e+r_b(1-\alpha_0)}$	$\dfrac{1}{1+(r_b/R_L)(1-\alpha_0)}$

A.6.17

Common base:

The equivalent T circuit for the common base configuration is shown in Fig. 6.17.1 and conventionally the currents flowing into the transistor are taken as positive.

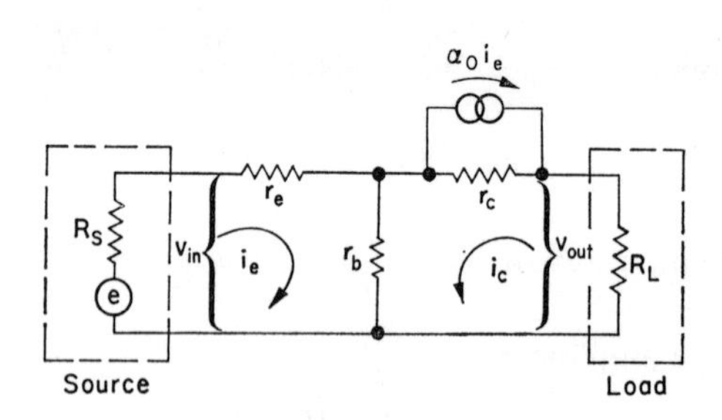

FIG. 6.17.1. Common base equivalent T circuit.

The circuit equations are:

$$v_{\text{in}} = i_e(r_e+r_b)+i_c r_b \quad (1)$$

$$= i_c(r_b+r_c+R_L)+i_e r_b+\alpha_0 i_e r_c, \quad (2)$$

$$v_{\text{out}} = -\, i_c R_L. \quad (3)$$

From (2)

$$\frac{i_c}{i_e} = -\frac{r_b+\alpha_0 r_c}{r_b+r_c+R_L}.$$

But $\alpha_0 r_c \gg r_b$.

$$\begin{aligned}
\text{Therefore current gain } &= \frac{i_c}{i_e} \\
&= -\frac{\alpha_0 r_c}{r_c+R_L} \\
&= -\frac{\alpha_0}{1+(R_L/r_c)} \\
&= -\alpha_0 \quad \text{if} \quad R_L \ll r_c.
\end{aligned}$$

The negative sign shows that i_c actually flows out of the transistor.

Eliminating i_c from (1),

$$v_{\text{in}} = i_e(r_e+r_b)-\alpha_0 r_b,$$

i.e.

$$Z_{\text{in}} = \frac{v_{\text{in}}}{i_e} = r_e+r_b(1-\alpha_0).$$

$$\begin{aligned}
\text{Then voltage gain } &= \frac{v_{\text{out}}}{v_{\text{in}}} \\
&= -\frac{i_c R_L}{i_e Z_{\text{in}}} \\
&= -\frac{\alpha_0 R_L}{r_e+r_b(1-\alpha_0)}.
\end{aligned}$$

The output impedance may be determined by applying a voltage to the output terminals as shown in Fig. 6.17.2. The source e.m.f. is short circuited.

Applying Kirchhoff's laws

$$v = i_c(r_c+r_b)+\alpha_0 i_e r_c+i_e r_b,$$
$$0 = i_e(r_e+r_b+R_S)+i_c r_b.$$

Eliminating i_e

$$v = i_c(r_c+r_b)+(\alpha_0 r_c+r_b)\left(-\frac{i_c r_b}{r_e+r_b+R_S}\right).$$

Since $\alpha_0 r \gg r_b$

$$v = i_c r_c - \frac{i_c \alpha_0 r_c r_b}{r_e+r_b+R_S},$$

therefore $$Z_{out} = \frac{v}{i_c} = r_c - \frac{\alpha_0 r_c r_b}{r_e+r_b+R_S}.$$

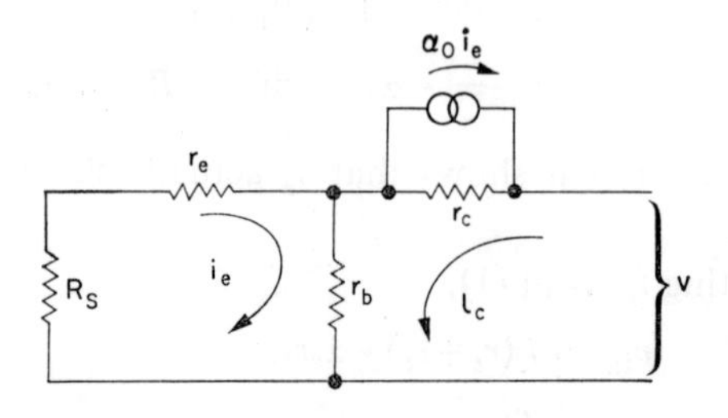

FIG. 6.17.2. Common base equivalent *T* circuit to determine output impedance.

The relations for the common emitter circuit are proved in *T.D.E.*, section 6.7.1, and it is left to the reader to prove the relations for the common collector circuit.

Q.6.18

A junction transistor has the following parameters:

$$r_b = 300\,\Omega, \qquad r_e = 20\,\Omega, \qquad r_c = 500\text{ k}\Omega, \qquad \alpha = 0{\cdot}97.$$

Draw an equivalent a.c. circuit for the transistor used in a common emitter arrangement, and determine the short-circuit current gain. Justify the method used. Small signal low frequency operation may be assumed.

Explain, with the aid of a circuit diagram, the method of biasing such a transistor by use of a potential divider across the collector–emitter supply voltage.

[B.Sc. 2, 1960]

A.6.18

The equivalent a.c. circuit for the common emitter arrangement is shown in Fig. 6.18.1. Applying Kirchhoff's laws to the output 1000,

$$i_c(5\times10^5+20)+20i_b+0{\cdot}97i_e\times5\times10^5 = 0,$$

$$i_e = -(i_b+i_c).$$

Therefore $i_c(5\times10^5+20)+20i_b-0{\cdot}97(i_b+i_c)\times5\times10^5 = 0,$

i.e. $$15020\,i_c-4{\cdot}85\times10^5\,i_b = 0.$$

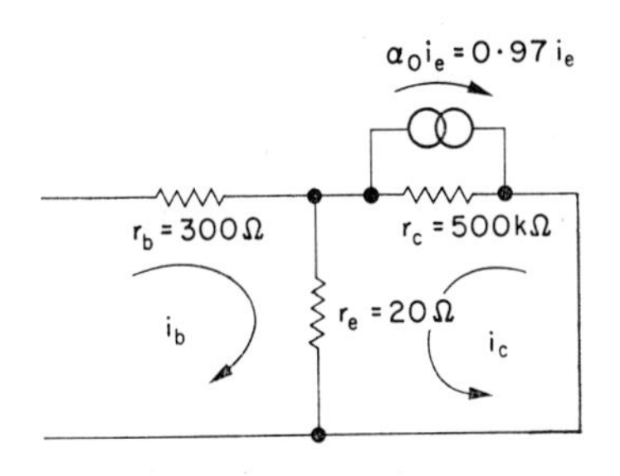

FIG. 6.18.1. Equivalent circuit for Q.6.18.

Therefore

$$\text{short circuit current gain} = \frac{4{\cdot}85\times10^5}{15020} = \underline{32{\cdot}3 \text{ times.}}$$

The second part of this question is covered in A.6.1.

Q.6.19

Prove that for the common base amplifier shown in Fig. 6.19.1 the output current falls by 3dB when

$$\omega = \frac{1}{C\{R_{\text{in}}+[R_L r_c/(R_L+r_c)]\}}$$

Assume a constant input current and a collector resistance much greater than the base resistance.

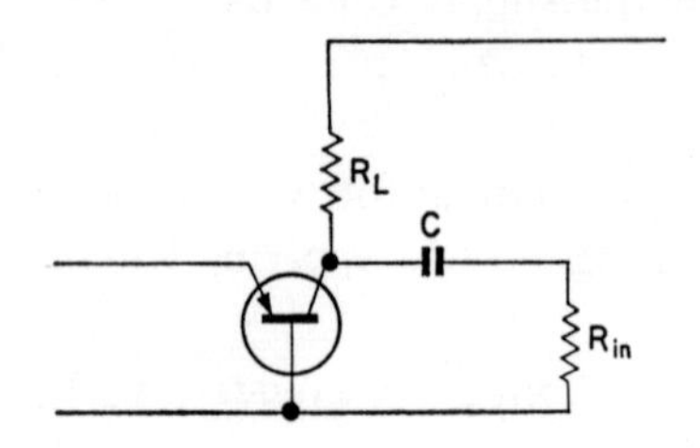

FIG. 6.19.1. Circuit for Q.6.19.

A.6.19

The equivalent circuit for the amplifier is shown in Fig. 6.19.2 and the circuit equations are:

$$i_c(r_b+r_c+R_L)+\alpha_0 i_e r_c+i_e r_b+i_{\text{out}} R_L = 0,$$

$$i_{\text{out}}\left(R_L+R_{\text{in}}+\frac{1}{j\omega C}\right)+i_c R_L = 0.$$

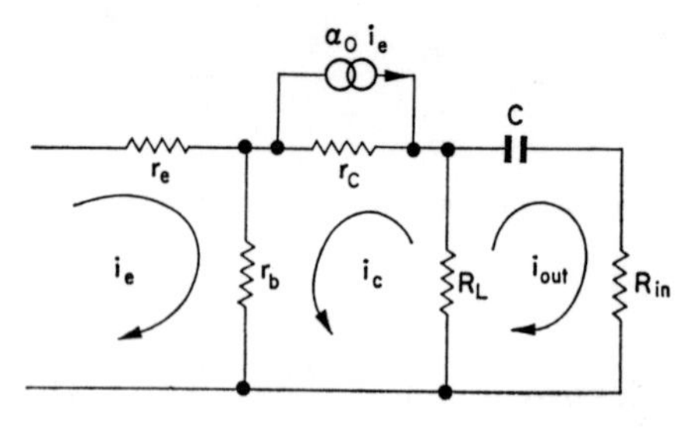

FIG. 6.19.2. Equivalent circuit for Q.6.19.

Eliminating i_c

$$-i_{\text{out}}\frac{[R_L+R_{\text{in}}+(1/j\omega C)]}{R_L}(R_L+r_b+r_c)+i_e(\alpha_0 r_c+r_b)$$
$$+i_{\text{out}}R_L = 0$$

or $i_{\text{out}}\{[R_L+R_{\text{in}}r(1/j\omega C)](R_L+r_b+r_c)-R_L^2\} = i_e(\alpha_0 r_c+r_b)R_L.$

Therefore overall current gain $= \dfrac{i_{out}}{i_e}$

$$= \frac{(\alpha_0 r_c + r_b)R_L}{R_L(r_b+r_c)+R_{in}(R_L+r_b+r_c)+(1/j\omega C)(R_L+r_b+r_c)}.$$

Since $\alpha_0 r_c \gg r_b$

Overall current gain

$$= \frac{\alpha_0 r_c R_L}{R_L r_c + R_{in}(R_L+r_b)+(1/j\omega C)(R_L+r_c)}$$

i.e. Overall current gain

$$= \frac{\alpha_0[R_L r_c/(R_L+r_c)]}{\{R_{in}+[R_L r_c/(R_L+r_c)]\}+1/j\omega C}.$$

The modulus of this is:

$$\frac{\alpha_0[r_c R_L/(R_L+r_c)]}{\sqrt{\{[R_{in}+(R_L r_c/R_L r_c)]^2+(1/\omega^2 C^2)\}}}$$

At h.f. this reduces to:

$$\frac{\alpha_0[r_c R_L/(R_L+r_c)]}{R_{in}+[r_c R_L/(R_L+r_c)]}.$$

Hence at the 3 dB point where the current gain falls to $1/\sqrt{2}$ of the h.f. value

$$\sqrt{\left\{\left(R_{in}+\frac{R_L r_c}{R_L+r_c}\right)^2+\frac{1}{\omega^2 C^2}\right\}} = \sqrt{2}\left(R_{in}+\frac{R_L r_c}{R_L+r_c}\right)$$

or

$$\frac{1}{\omega^2 C^2} = \left(R_{in}+\frac{R_L r_c}{R_L+r_c}\right)^2,$$

i.e.

$$\omega = 1/C\left(R_{in}+\frac{R_L r_c}{R_L+r_c}\right).$$

Q.6.20

Calculate after deriving the necessary formula the input resistance of a common collector transistor stage with emitter load resistance of 5 kΩ. The working point is such that

$$r_e = 40\ \Omega, \qquad r_b = 500\ \Omega, \qquad r_c = 1\ \text{M}\Omega, \qquad \alpha_0 = 0{\cdot}98.$$

Show an arrangement for obtaining the necessary base bias current that will not result in an excessive lowering of the input impedance.

[I.E.E., Nov. 1958]

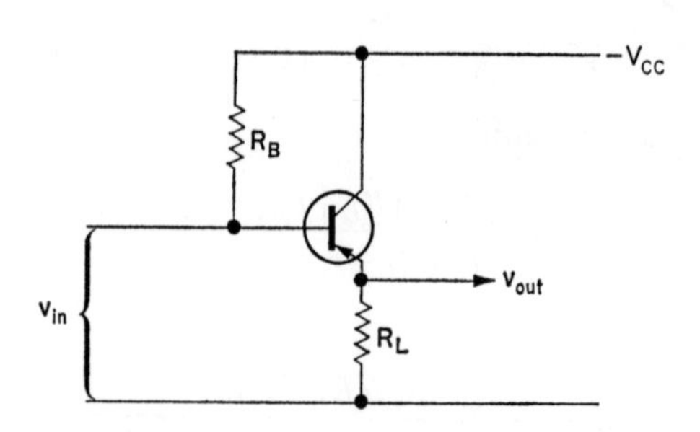

FIG. 6.20.1. Typical common collector stage.

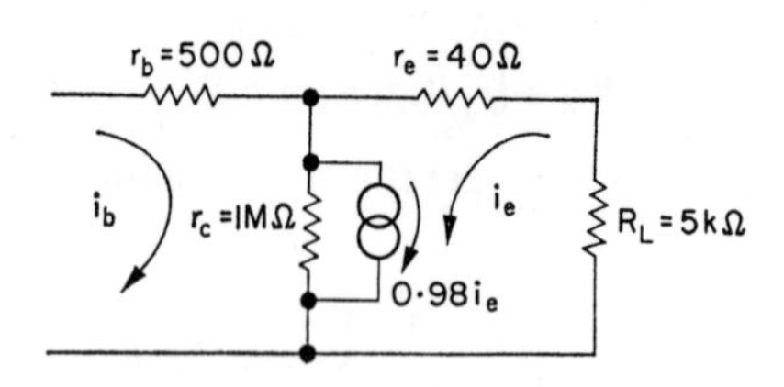

FIG. 6.20.2. Equivalent circuit for Q.6.20.

A.6.20

A typical common collector transistor stage is shown in Fig. 6.20.1. The equivalent circuit is shown in Fig. 6.20.2 and using Kirchhoff's laws the following equations may be derived:

$$v_{in} = i_b(500+10^6) - 0{\cdot}98 i_e \times 10^6 + i_e \times 10^6,$$
$$0 = i_e(10^6 + 5\times 10^3 + 40) - 0{\cdot}98 i_e \times 10^6 + i_b \times 10^6.$$

Eliminating i_e

$$v_{\text{in}} = i_b(500+10^6)-2\times10^4\left(\frac{i_b\times10^6}{2\times10^4+5\times10^3+40}\right).$$

Therefore
$$Z_{\text{in}} = \frac{v_{\text{in}}}{i_b}$$
$$= 500+10^6-0{\cdot}8\times10^6$$
$$= \underline{200{\cdot}5\ \text{k}\Omega}.$$

Alternatively the expression for the input impedance of a common collector stage shown in Q.6.17 may be derived and the values for r_e, r_b, r_c, R_L and α_0 substituted.

Figure 6.20.1 shows that the base bias resistor R_B shunts the input circuit. Due to the relatively high voltage developed across R_L the d.c. voltage drop across R_B may be one or two volts. Hence, the value of R_B may be considerably less than the input impedance of the stage.

One method of counteracting this is to feed the base from a separate supply through a high resistance. If only one supply is available a "bootstrap" circuit may be used. This is shown in Fig. 6.20.3 and although the d.c. resistances in the base circuit

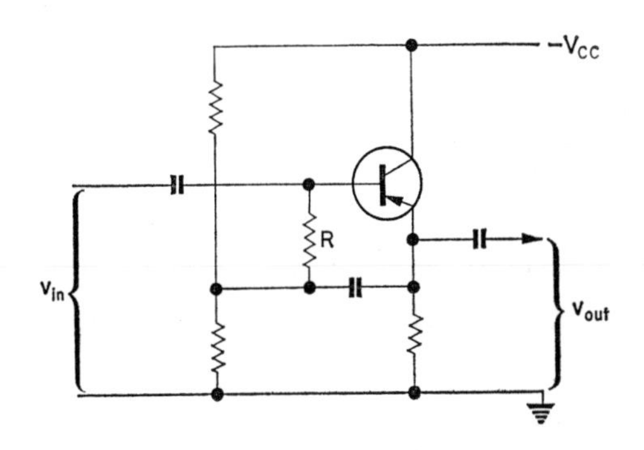

FIG. 6.20.3. High input impedance transistor circuit.

are relatively low the a.c. input impedance is very high since the "apparent" value of R is much greater than the physical value.

Q.6.21

The parameters of a transistor used as a grounded emitter a.f. amplifier are:

$$r_e = 50\,\Omega, \qquad r_b = 500\,\Omega, \qquad r_c = 1\,\text{M}\Omega, \qquad \alpha_0 = 0{\cdot}98.$$

Determine from first principles the input and output resistances of the stage, assuming a load resistance of 8 kΩ and a generator resistance of 850 Ω.

[H.N.C. 3, 1963]

A.6.21

The equivalent circuit for the grounded emitter amplifier is shown in Fig. 6.21.1.

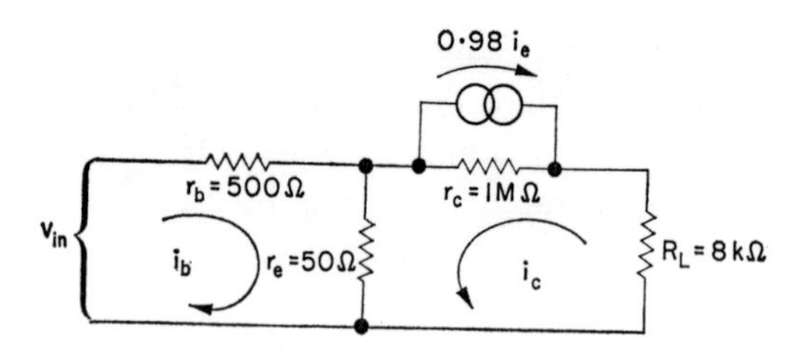

FIG. 6.21.1. Common emitter equivalent circuit to determine output impedance.

The circuit equations are:

$$v_{\text{in}} = i_b 550 + i_c\, 50, \tag{1}$$

$$0 = i_c(10^6 + 8\times 10^3 + 50) + i_b 50 + 0{\cdot}98 \times i_e \times 10^6, \tag{2}$$

$$i_e = -(i_b + i_c).$$

From (2) and (3)

$$0 = i_c(2\times 10^4 + 8\times 10^3 + 50) + i_b(50 - 0{\cdot}98 \times 10^6)$$

or $$i_c = \frac{0{\cdot}98\times10^6}{2{\cdot}8\times10^4}\, i_b$$
$$= 35 i_b.$$

From (1) $$v_{\text{in}} = i_b\, 550 + i_b\, 1750.$$

Therefore $$Z_{\text{in}} = \frac{v_{\text{in}}}{i_b}$$
$$= \underline{2300\ \Omega.}$$

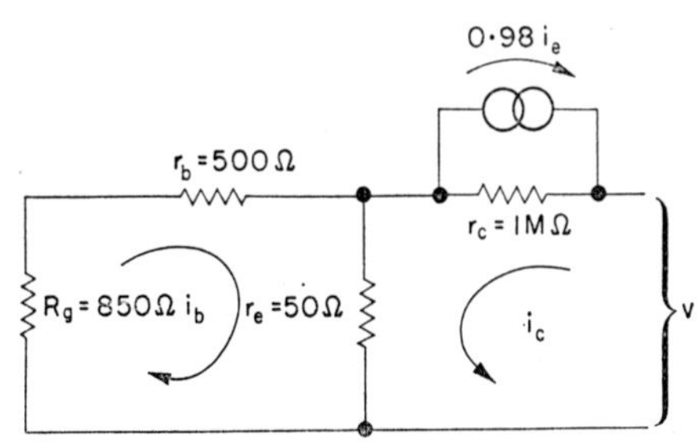

FIG. 6.21.2. Equivalent circuit of common emitter amplifier.

Figure 6.21.2 shows the amplifier with a voltage v applied to the output terminals. The equations are:

$$v = i_c(10^6+50)+i_b 50+0{\cdot}98\times10^6 i_e, \quad (1)$$
$$0 = i_b(500+50+850)+i_c\, 50, \quad (2)$$
$$i_e = -\,(i_b+i_c). \quad (3)$$

From (2) $$i_b = -\frac{i_c\, 50}{1400}$$
$$= -\frac{i_c}{28}.$$

From (1) and (3) $$v = i_c(10^6+50)+i_b 50-0{\cdot}98\times10^6(i_b+i_c)$$
$$= i_c\times 20050 - i_b 0{\cdot}98\times10^6$$
$$= i_c\times 20050 + i_c\,\frac{0{\cdot}98\times10^6}{28}.$$

Therefore $$Z_{\text{out}} = \frac{v}{i_c} = 20050+35000$$

$$= \underline{55\ \text{k}\Omega.}$$

Q.6.22

The hybrid parameters for a transistor used in the common emitter configuration are:

$$h_{ie} = 1{\cdot}5\ \text{k}\Omega, \quad h_{re} = 10^{-4}, \quad h_{fe} = 70, \quad h_{0e} = 100\ \mu\text{mho}$$

The transistor has a load resistor of 1 kΩ in the collector lead, and is supplied from a signal source of resistance 800 Ω.

Calculate (a) the input resistance, (b) the output resistance, (c) the voltage gain, and (d) the current gain for the stage. Draw a circuit for the amplifier showing bias arrangements.

[I.E.E., Nov. 1963]

A.6.22

The equivalent circuit using the h parameters is shown in Fig. 6.22.1. From this circuit

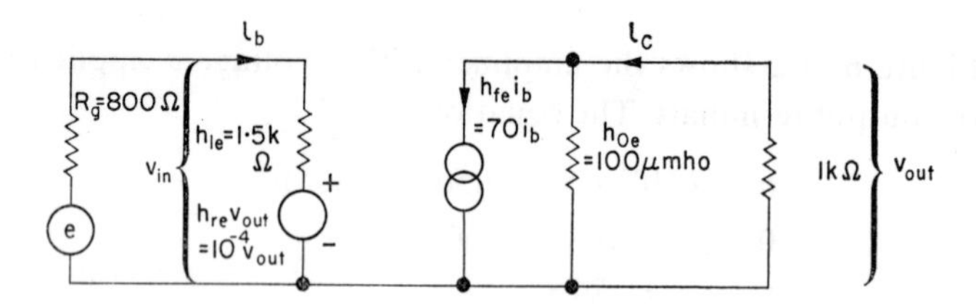

FIG. 6.22.1. Equivalent circuit using h parameters for Q.6.22.

$$i_c = v_{\text{out}} h_{0e} + h_{fe} i_b$$

$$= v_{\text{out}} \times 10^{-4} + 70 i_b, \tag{1}$$

$$v_{\text{in}} = i_b \times h_{ie} + v_{\text{out}} h_{re}$$

$$= i_b \times 1{\cdot}5 \times 10^3 + v_{\text{out}} \times 10^{-4}, \tag{2}$$

$$v_{\text{in}} = E - 800 i_b, \tag{3}$$

$$v_{\text{out}} = -i_c \times 10^3. \tag{4}$$

From (1) and (4) $\quad i_c = -0{\cdot}1 i_c + 70 i_b$,

i.e. $\quad i_c = 63{\cdot}6\, i_b$.

Therefore current gain = 63·6 times

and $\quad v_{out} = -63{\cdot}6 \times 10^3 i_b$.

Hence from (2) $\quad v_{in} = i_b \times 1{\cdot}5 \times 10^3 - 6{\cdot}36 i_b$.

Therefore
$$Z_{in} = \frac{v_{in}}{i_b}$$
$$= \underline{1{\cdot}5 \text{ k}\Omega.}$$

Therefore voltage gain
$$= \frac{v_{out}}{v_{in}}$$
$$= \frac{-63{\cdot}6 \times 10^3 i_b}{1500 i_b}$$
$$= -42{\cdot}4 \text{ times.}$$

This is the voltage gain of the transistor itself.

The overall voltage gain
$$= \frac{v_{out}}{E}$$
$$= \frac{-63{\cdot}6 \times 10^3 i_b}{2300 i_b}$$
$$= \underline{-27{\cdot}7 \text{ times.}}$$

The output impedance may be derived by applying a voltage V to the output terminals and short circuiting the source e.m.f.

Then
$$i_c = v \times 10^{-4} + 70 i_b,$$
$$0 = i_b \times 2300 + v \times 10^{-4}.$$

Therefore
$$i_c = v \times 10^{-4} + 70\left(\frac{-10^{-4} v}{2300}\right),$$

i.e.
$$v = \frac{2300 i_c}{10^{-4} \times 2230}.$$

Therefore
$$Z_{out} = \frac{v}{i_c} = \underline{10{\cdot}3 \text{ k}\Omega.}$$

The circuit for the amplifier is shown in Fig. 6.22.2. Potential divider and emitter bias resistors are included, but no values are given since there is insufficient information about the d.c. characteristics of the transistor.

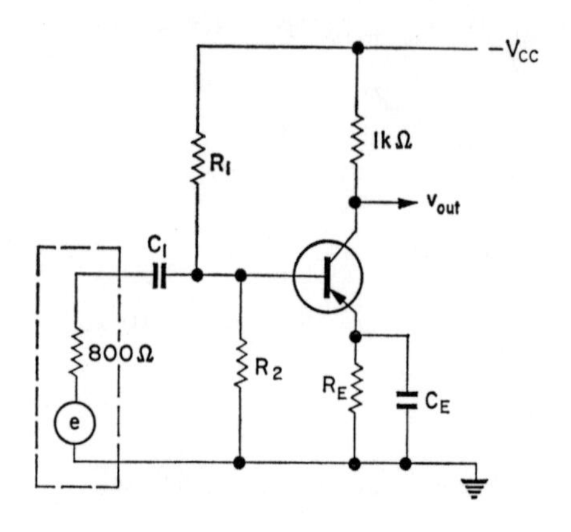

FIG. 6.22.2. Circuit for Q.6.22.

Q.6.23

Describe one method for the measurement of each of the hybrid parameters of a junction transistor in the common emitter connection.

The hybrid parameters of a given transistor connected with common emitter are as follows:

$$h_{11} = 400\ \Omega, \qquad h_{21} = 50,$$
$$h_{12} = 0{\cdot}5\times 10^{-3}, \qquad h_{22} = 25\times 10^{-6}\ \text{mho}.$$

If this transistor is used in an amplifying circuit with a collector load of 1 kΩ, calculate the alternating voltage at the collector if the base is supplied from a generator having an internal resistance of 600 Ω and an open circuit voltage of 0·01 V r.m.s. The effect of the d.c. bias resistance may be neglected.

[B.Sc. 3, 1959]

A.6.23

The first part of this question is answered in A.4.3.

The example is very similar to Q.6.22 and using the equations

$$
\begin{aligned}
i_c &= v_{out}h_{22} + i_b h_{21}, \\
v_{in} &= i_b h_{11} + v_{out}h_{12}, \\
v_{in} &= E - i_b R_g, \\
v_{out} &= -i_c R_L,
\end{aligned}
$$

it may be shown that $v_{out} = \underline{0 \cdot 5 \text{ V r.m.s.}}$

CHAPTER 7

Power Amplifiers

Q.7.1. Low power, Class A amplifier with resistive load.
Q.7.2. Low power amplifier with transformer coupled load.
Q.7.3. Load line calculation for maximum undistorted power.
Q.7.4. Push–pull amplifier with complementary transistor pair.
Q.7.5. Power output from a common emitter amplifier.
Q.7.6. Class A push–pull amplifier, calculation of harmonic distortion.
Q.7.7. Definitions of Class A, B and C, with calculations for a Class C amplifier.
Q.7.8. Calculations for a transformer coupled Class A power amplifier.
Q.7.9. Class C amplifier, with further calculations.

Q.7.1

Explain the difference between the transistor current gains α and β and their relationship to one another.

The collector current–collector voltage characteristics of a junction transistor in the common emitter configuration, with the base current as a parameter, are as shown in Table 7.1.

TABLE 7.1

V (volts)	$-I_C$ (mA)		
	$-I = -40\ \mu A$	$80\ \mu A$	$120\ \mu A$
0·5	0·6	1·8	2·4
2	0·8	2·8	4·8
6	1·0	3·0	5·0
12	1·3	3·3	5·3

The transistor is to be used as a common emitter amplifier with a battery supply of 12 V and a collector load resistance of 2 kΩ. The quiescent operating point is $V_C = -6$ V, $I_C = -3{\cdot}0$ mA.

If the base current is changed 40 μA by a peak input signal of 100 mV, determine the current gain, the voltage gain and the power gain. Express the power gain in decibels.

[H.N.D. 2, 1963]

A.7.1

The first part of this question is answered in A.3.4.

The collector characteristics of the given transistor are plotted in Fig. 7.1.1. The "load line" for a resistance of 2 kΩ, with a battery supply of 12 V, is also drawn across these characteristics.

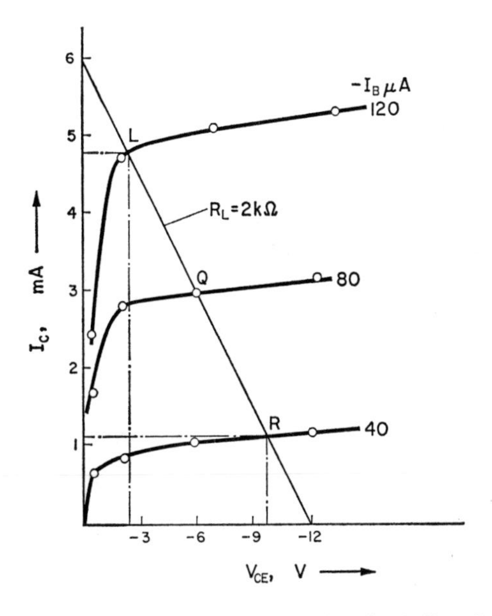

FIG. 7.1.1. Transistor characteristics, with load line for 2 kΩ of A.7.1.

The operating point corresponding to $V_{CE} = -6$ V and $I_C = -3$ mA is indicated by Q on the graph. The gains for a base current change (peak to zero) of 40 μA may be read off from the points L and R at the intersections of the load line, and the characteristics for $I_B = -120$ μA and $I_B = -40$ μA.

The collector current swing is

$$4{\cdot}9 - 1{\cdot}2 = 3{\cdot}7 \text{ mA}$$

and the collector voltage swing is

$$10 - 2{\cdot}8 = 7{\cdot}2 \text{ V}.$$

Hence

$$\text{current gain} = \frac{3{\cdot}7 \text{ mA}}{2 \times 40\ \mu\text{A}} = \underline{46 \text{ times}}$$

and

$$\text{voltage gain} = \frac{7{\cdot}2 \text{ V}}{2 \times 100 \text{ mV}} = \underline{36 \text{ times.}}$$

Therefore

$$\text{power gain} = 36 \times 46 \simeq \underline{1600 \text{ times.}}$$

or, in decibels,

$$\text{power gain} = 10 \log_{10}(1600) = \underline{32 \text{ dB.}}$$

Q.7.2

The collector characteristics of a certain transistor, when used in the grounded emitter configuration, may be assumed linear between the points given in Table 7.2.

TABLE 7.2

I_B (μA)	I_C (mA)	
	$V_C = -0{\cdot}4$ V	$V_C = -10$ V
20	0·9	1·5
60	2·8	4·2
100	4·8	6·6
140	6·6	9·8

By drawing the output characteristics and superimposing the load line for the low power a.f. amplifier of Fig. 7.2.1 estimate the following:

(a) mean input power;
(b) a.c. output power;
(c) total power dissipation of stage;
(d) efficiency;
(e) current gain } between base and collector;
(f) voltage gain }
(g) power gain.

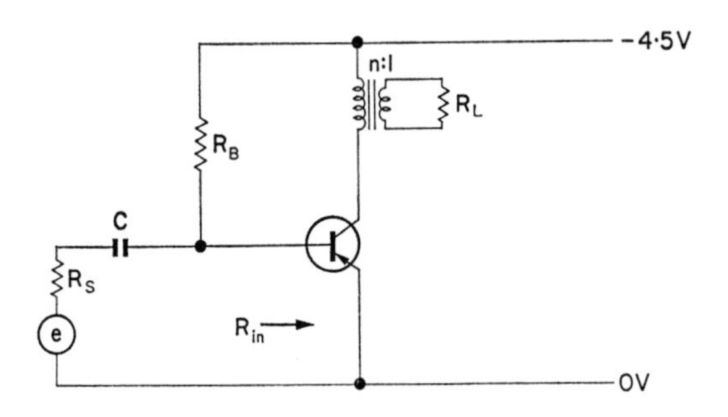

FIG. 7.2.1. Transformer coupled amplifier circuit for Q.7.2.— $e = 100 \sin \omega t$ mV; $R_S = 1000\ \Omega$; $R_B = 72\ \text{k}\Omega$; $R_L = 20\ \Omega$; $n = 10$; a.c. input resistance $R_{in} = 1500\ \Omega$; base-emitter d.c. potential $= -0{\cdot}18$ V; reactance of C negligible at the working frequency.

A.7.2

The output characteristics given in Table 7.2 are plotted in Fig. 7.2.2. With a transformer having a step-down ratio of 10:1, the effective load resistance is $R_L = (10)^2 \times 20 = 2000\ \Omega$, and this determines the slope of the load line.

For a base to emitter d.c. potential of 0·18 V the direct current at the base is

$$I_B = \frac{4{\cdot}5 - 0{\cdot}18}{72 \times 10^3}\ \text{A} = 60\ \mu\text{A}.$$

The operating point Q must therefore be at $I_B = 60\ \mu A$ and $V_{CE} = -4{\cdot}5$ V, and the load line is therefore drawn through this point, with a slope of 1/2000. The available peak to zero input voltage is

$$v_{in} = \frac{100 \times 1500}{1500 + 1000} \text{ mV} = 60 \text{ mV}$$

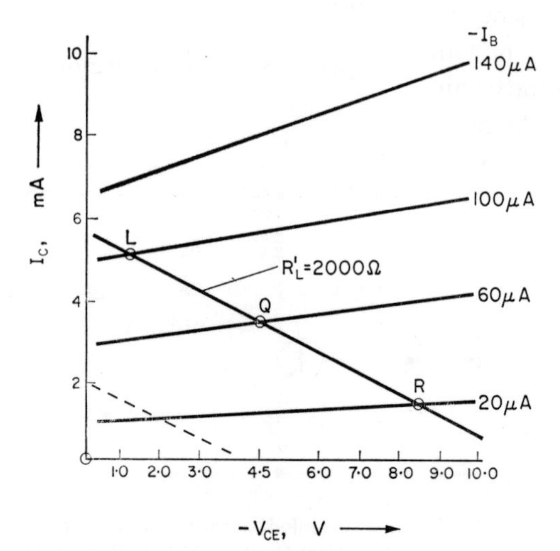

FIG. 7.2.2. Characteristics of Table 7.2, with effective load line for 2000 Ω.

and hence the peak to zero input current at the base is

$$\frac{60}{1500} \text{ mA} = 40\ \mu\text{A}.$$

Therefore the mean input power is

$$P_{in} = \frac{(60 \times 10^{-3})(40 \times 10^{-6})}{2} \text{ W} = 1{\cdot}2\ \mu\text{W}.$$

By horizontal and vertical projections, from points L, Q and R in Fig. 7.2.2, we can see that the peak to zero output current and voltage are approximately

$$i_{\text{out}} = \frac{5-1\cdot4}{2} = 1\cdot8 \text{ mA}$$

and
$$v_{\text{out}} = \frac{8\cdot3-1\cdot5}{2} = 3\cdot4 \text{ V.}$$

$$\text{Hence, current gain} = \frac{1800}{40} = 45,$$

$$\text{voltage gain} = \frac{3400}{60} = 56,$$

$$\text{and power gain} = 45\times56 \simeq 2500.$$

Therefore the a.c. output power $= 2500\times1\cdot2\times10^{-6}$ W

or
$$P_{\text{out}} = 3 \text{ mW.}$$

The d.c. power available at the collector is given by the no signal conditions at the operating point, so

$$P_c = (4\cdot5\times3\cdot3) \text{ mW} \simeq 15 \text{ mW.}$$

The collector dissipation is then $15-3 = 12$ mW and the efficiency is

$$(3/15)\ 100\% = 20\%.$$

The d.c. power dissipated in the base input circuit is low enough to be ignored in this calculation.

To summarize these results, in the same order as in the question:

(a) 1·2 μW. (e) 45.
(b) 3 mW. (f) 56.
(c) 12 mW. (g) 2500.
(d) 20%.

Q.7.3

A *pnp* germanium transistor is used in the common emitter connection with an ideal 3 : 1 step-down transformer in the collector circuit. The operating point is determined by a resistor connected between the base and the negative supply.

If the supply is 16 V and the collector dissipation is not to exceed 250 mW, determine graphically a suitable operating point to give approximately maximum "undistorted" output

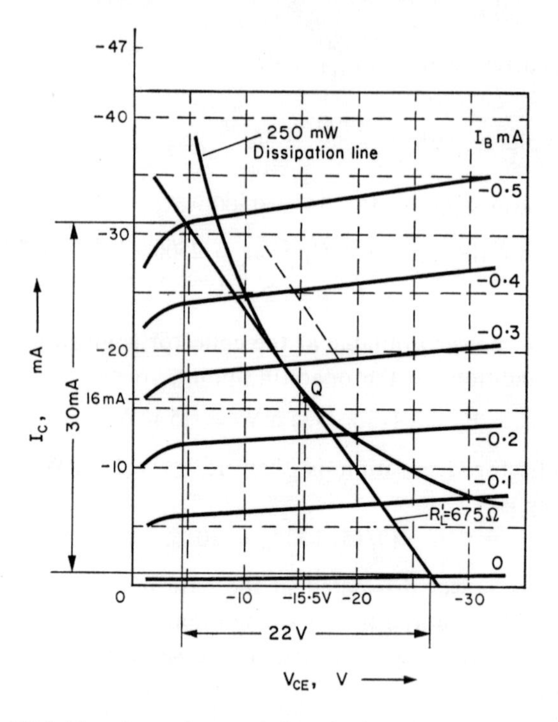

FIG. 7.3.1. Transistor characteristics with load line and maximum dissipation curve for Q.7.3.

in a load resistor of 75 Ω connected across the secondary of the transformer. Calculate the power in the load resistor and the value of the bias resistor.

(Transistor output characteristics provided in Fig. 7.3.1.)

[I.E.E., Nov. 1962]

A.7.3

The curve corresponding to a maximum collector dissipation of 250 mW is drawn across the transistor characteristics of Fig. 7.3.1. The effective a.c. load resistance at the collector is

$$R'_L = (75)\,(3)^2 = 675\,\Omega.$$

For maximum output power, under these conditions, the load line should approximate to a tangent to the maximum dissipation curve, and have a slope of 1/675 mho.

Under a.c. conditions the base current variation should not be large enough to reduce the collector voltage below the "knee" of the characteristics. A suitable load line to obtain maximum "undistorted" power is shown superimposed on the characteristics.

This gives a maximum voltage swing of 22 V peak to peak and a maximum current swing of 30 mA peak to peak about an operating point given by $V_{CE} = -15{\cdot}5$ V, $I_C = 16$ mA and $I_B = 0{\cdot}25$ mA.

The power developed in the load resistor is then

$$\frac{22}{2\sqrt{2}} \times \frac{30}{2\sqrt{2}} = \underline{82{\cdot}5 \text{ mV.}}$$

Since the supply voltage is 16 V, 0·5 V must be dropped in the circuit. In practice, some of this would be dropped in transformer primary winding. Assuming a perfect transformer the resistance required is 0·5 V/16 mA = 30 Ω and this could easily be connected in the emitter circuit. If it is left undecoupled some negative feedback would be introduced which would have a beneficial effect. This assumes that the input voltage is of sufficient amplitude to give the necessary base current variation.

The base bias resistor must be of such a magnitude that the d.c. base current is 0·25 mA. Allowing for the 0·5 V drop across the emitter resistor and say 0·3 V d.c. across the base emitter diode this becomes 15·2 V/0·25 mA = 60kΩ.

The complete circuit is shown in Fig. 7.3.2.

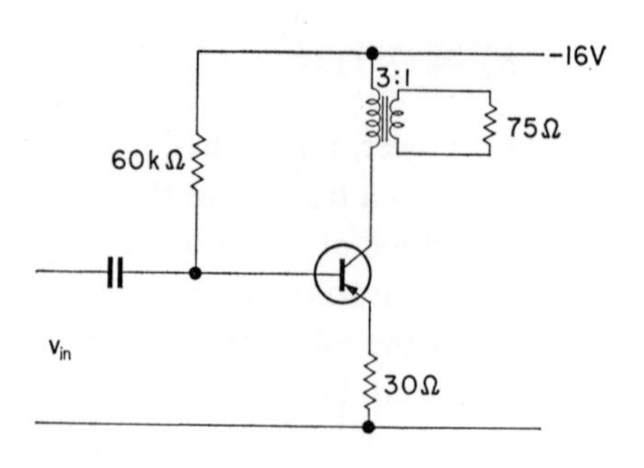

FIG. 7.3.2. Circuit for Q.7.3.

Q.7.4

Discuss the advantages of a push–pull audio-frequency amplifier using a pair of complementary transistors. Also point out the problems associated with using *pnp* and *npn* transistors together in this way.

A push–pull common emitter amplifier stage has complementary transistors, working into a resistive load, with linear characteristics and correct d.c. bias conditions. However, the current gains of the two transistors are 50 and 40 respectively. Show how the output waveform may be derived from the d.c. characteristics of the transistors and estimate the percentage harmonic distortion if this may be assumed to be only second harmonic.

[H.N.C. 3, 1963]

A.7.4

The primary advantage of the use of a pair of complementary transistors is to eliminate the output transformer in a push–pull power amplifier. The complementary amplifier can thus drive into a resistive load whilst retaining the high efficiency of near Class B operation. As the transformer is often the most bulky and expensive item in power amplifiers, operating at audio frequencies, the complementary pair of transistors can lead to a reduction in both size and cost of the amplifier. In addition the frequency response of the amplifier may be improved. However, it is essential that the pair of transistors (one *pnp* and the other *npn*) should be matched in all characteristics, over the whole working range of frequency and temperature. It is extremely difficult to manufacture matched high power complementary transistors, and therefore it is usual to put the complementary pair in the driving stage and not in a high power output stage.

In most cases it is uneconomical to match pairs of transistors within closer than 5% and divergences of greater than 20% are to be expected at the extremes of frequency range, d.c. conditions or temperature.

Typical d.c. output characteristics for a *pnp/npn* pair are plotted in Fig. 7.4.1. These curves are drawn on the assumption that the *pnp* transistor has a mean current gain of 50 and *npn* transistor a gain of 40. The output waveform from the push–pull pair is derived by considering equal positive and negative swings of base current and finding the corresponding collector swings from the characteristics. A typical load line is drawn through an assumed supply voltage of 12 V (positive for the *npn* and negative for the *pnp* transistor). In practice the slope of this load line would depend on the load resistance.

For the assumed load line an input current swing of 0·75 mA peak gives rise to an output current variation of +31·5 mA to −39 mA.

If we assume the distortion is due only to second harmonic distortion, one peak is increased by an amount equal to the peak value of the second harmonic and the other peak reduced. This is shown in Fig. 7.4.2.

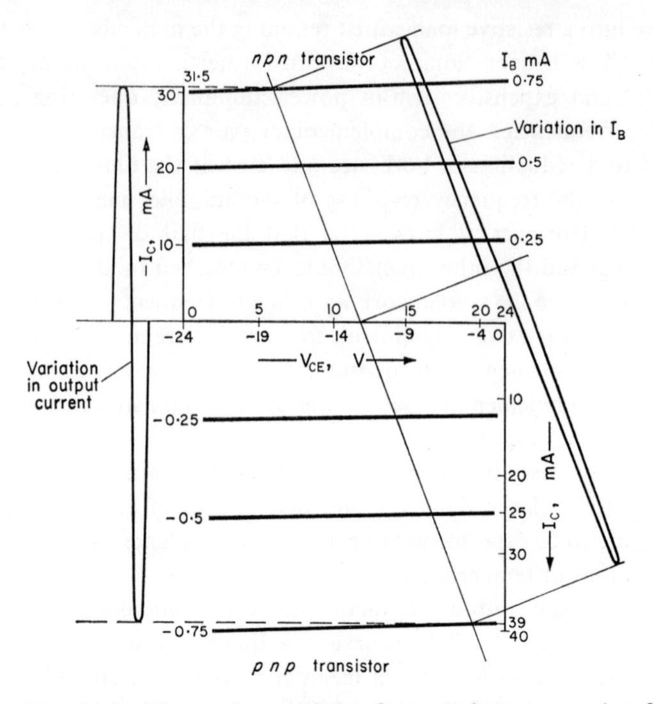

FIG. 7.4.1. Output characteristics for complementary pair of transistors.

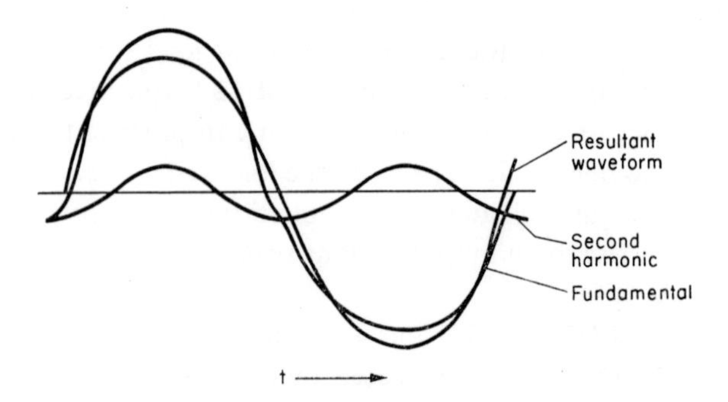

FIG. 7.4.2. Waveform showing the effect of second harmonic distortion.

Hence, the peak to peak value of the second harmonic is $39-31{\cdot}5 = 7{\cdot}5$ mA.

The peak to peak value of the fundamental is then the sum of the peak values, i.e. $39+31{\cdot}5 = 70{\cdot}5$ mA.

The percentage second harmonic distortion is then

$$7{\cdot}5/70{\cdot}5\times 100 = 10{\cdot}6\%.$$

Q.7.5

A *pnp* transistor has the following common base parameters:

$$r_e = 50\ \Omega, \qquad r_b = 500\ \Omega, \qquad r_c = 500\ \text{k}\Omega, \qquad \alpha = 0{\cdot}96.$$

Derive an equivalent circuit to represent the behaviour of this transistor when used in the common emitter configuration.

The transistor is used as a common emitter amplifier, the input being obtained from a generator of resistance 600 Ω and e.m.f. 150 mV (r.m.s.) Obtain an expression for the power delivered to a resistance R_L connected between collector and emitter. For what value of R_L is this power a maximum and what is the numerical value of this maximum power?

[B.Sc. 3, 1962]

A.7.5

The solution of this question illustrates the use of the constant voltage form of equivalent T circuit. The equivalent circuit using the current generator could equally well be used.

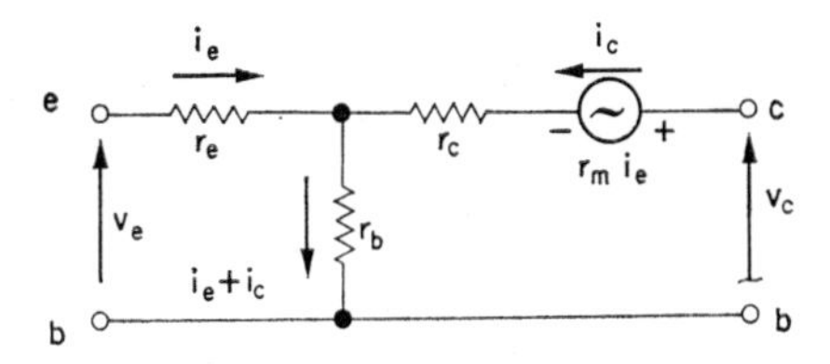

FIG. 7.5.1. Common base equivalent circuit for transistor.

The common base equivalent circuit is shown in Fig. 7.5.1. This can be re-drawn for the common emitter configuration as shown in Fig. 7.5.2.

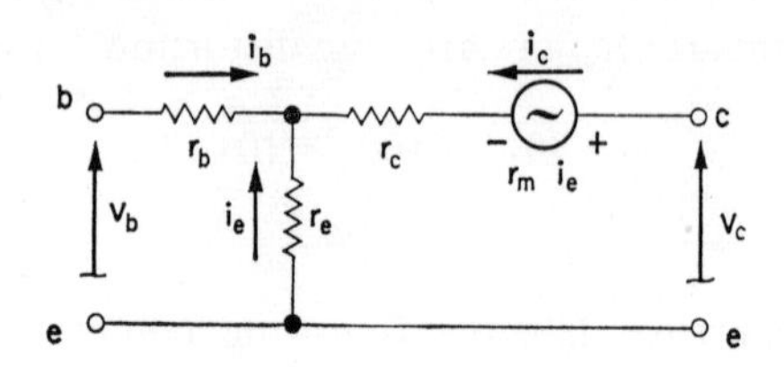

FIG. 7.5.2. Equivalent circuit redrawn for common emitter.

The equations for this circuit are:

$$v_b = i_b r_b - i_e r_e, \tag{1}$$

$$v_c = i_c r_c + i_e r_m - i_e r_e, \tag{2}$$

$$0 = i_e + i_c + i_b. \tag{3}$$

Eliminating i_e,

$$v_b = i_b(r_b + r_e) + i_c r_e, \tag{4}$$

$$v = i_c(r_c - r_m + r_e) - i_b r_m + i_b r_c. \tag{5}$$

Now $\qquad r_m = \alpha r_c.$

Therefore $\qquad v_c = i_c[r_c(1-\alpha) + r_e] + i_b r_e - i_b r_m. \tag{6}$

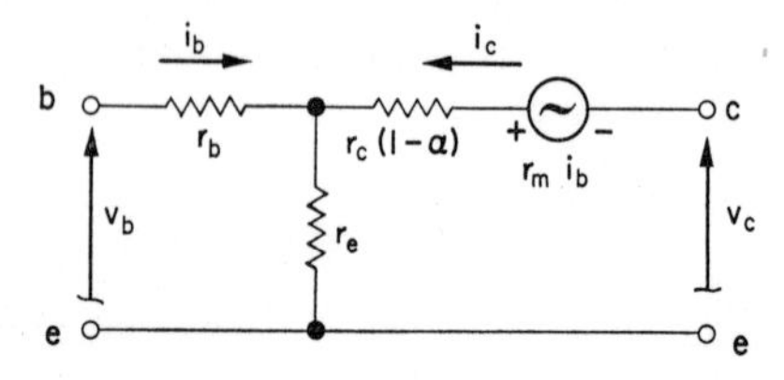

FIG. 7.5.3. Modified equivalent circuit for A.7.5.

The circuit represented by this equation is shown in Fig. 7.5.3.

Substituting the values given we obtain:

$$r_c(1-\alpha) = 500\,(1-0{\cdot}96)\times 10^3 = 20\ \text{k}\Omega,$$

$$r_m = \alpha r_c = 0{\cdot}96\times 500\times 10^3 = 480\ \text{k}\Omega.$$

The equivalent circuit for the amplifier can now be drawn as shown in Fig. 7.5.4.

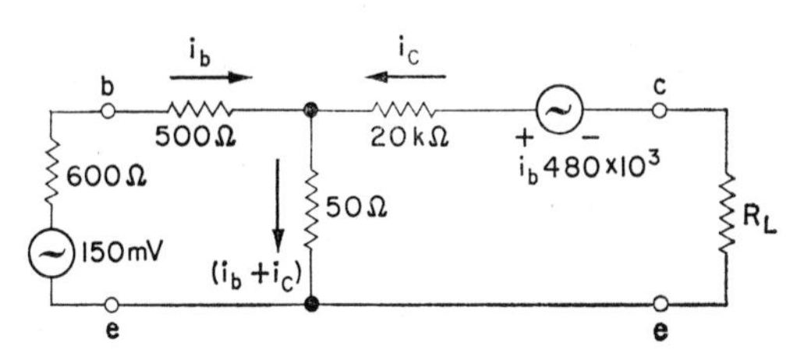

FIG. 7.5.4. Equivalent circuit for amplifier of A.7.5.

The equations are:

$$0{\cdot}15 = i_b(600+500+50)+i_c\,50.$$

Therefore $$0{\cdot}15\times 1150 i_b+50 i_c, \qquad (1)$$

$$i_b 480\times 10^3 = i_c(20\times 10^3+50+R_L)+50 i_b,$$

$$i_b(480\times 10^3-50) = i_c(20\times 10^3+50+R_L).$$

Therefore $$480\times 10^3\, i_b = (20\times 10^3+R_L)i_c. \qquad (2)$$

From (1) $$i_b = \frac{0{\cdot}15-50 i_c}{1150}.$$

Therefore $$480\times 10^3\left(\frac{0{\cdot}15-50 i_c}{1150}\right) = (20\times 10^3+R_L)i_c,$$

$$62{\cdot}6-20{\cdot}85\times 10^3 i_c = (20\times 10^3+R_L)i_c,$$

$$i_c = \frac{62{\cdot}6}{40{\cdot}85\times 10^3+R_L}\ \text{amps.}$$

Power in $$R_L = \left(\frac{62{\cdot}6}{40{\cdot}85\times 10^3-R_L}\right)^2 R_L\ \text{watts.}$$

By differentiating and equating to zero it can be shown that this will be a maximum when $R_L = 40{\cdot}85\ \text{k}\Omega \triangleq 41\ \text{k}\Omega$.

Then
$$P_{max} = \left(\frac{62{\cdot}6}{81{\cdot}8\times10^3}\right)^2 41\times10^3$$
$$= \underline{23 \text{ mW.}}$$

Q.7.6

Explain briefly the function of and the requirements for each component in a single-stage, push–pull, Class A amplifier employing two *npn* transistors.

Assume an ideal transformer, with a resistive load, and transistor characteristics given by

$$i_{c_1} = 300(v_{be_1}-0{\cdot}1)+10(v_{be_1}-0{\cdot}1)^2$$

and

$$i_{c_2} = 250(v_{be_2}-0{\cdot}11)+9(v_{be_2}-0{\cdot}11)^2,$$

where the collector currents are in milliamperes and the base to emitter voltages in volts.

The base bias voltage is −350 mV at each transistor and the peak sinusoidal signal voltage is 100 mV to each base. Determine the percentage second harmonic distortion at the output.

What is the meaning of this answer?

[H.N.D. 3, 1964]

A.7.6

The circuit diagram for a push–pull, Class A amplifier is given as Fig. 7.6.1. The variable resistor R_1 is adjusted such that for a given value of R_2 both transistors are conducting over the whole range of input voltage.

The input transformer must be accurately centre tapped to ensure that each transistor receives exactly one half of the input waveform, in antiphase.

In the same way the primary of the output transformer must have an exact centre tap, to minimize the effect of direct current flowing through the two halves of the winding and the effect of even harmonics produced in the transistors.

The emitter resistor R_E provides a source of d.c. feedback which reduces the danger of thermal runaway at elevated temperatures. It is usual for the d.c. volt drop across R_E to be about 10% of the supply voltage, since the emitter resistor is common to both transistors, which are driven with antiphase voltages from the input transformer. There is no need for this resistor to be decoupled at the signal frequency.

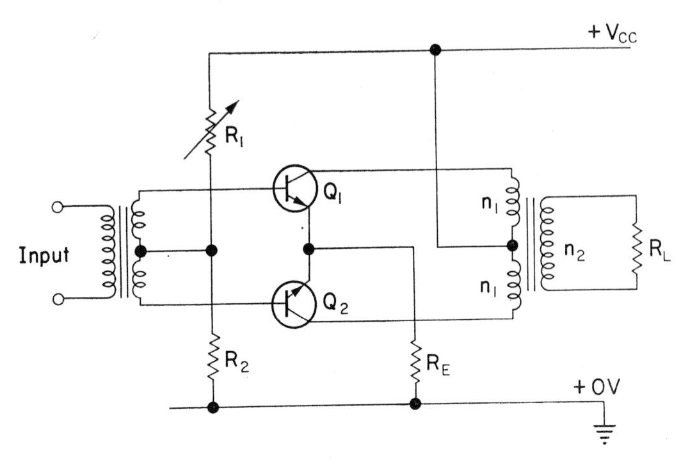

FIG. 7.6.1. Push–pull, Class A amplifier, with *npn* transistors.

The output voltage from the amplifier is proportional to the instantaneous difference between the collector currents. That is

$$v_{\text{out}} = K(i_{c_1} - i_{c_2}),$$

$$v_{be_1} = (350 + 100 \sin \omega t)\text{ mV},$$

and

$$v_{be_2} = (350 - 100 \sin \omega t)\text{ mV}$$

at any instant.

Hence,

$$i_{c_1} = 300(0{\cdot}25 + 0{\cdot}1 \sin \omega t) + 10(0{\cdot}25 + 0{\cdot}15 \sin \omega t)^2$$

and

$$i_{c_2} = 250(0{\cdot}24 - 0{\cdot}1 \sin \omega t) + 9(0{\cdot}24 - 0{\cdot}1 \sin \omega t)^2$$

or $i_{c_1} = 75+30 \sin \omega t+0{\cdot}625+0{\cdot}1 \sin^2 \omega t+0{\cdot}5 \sin \omega t$

and $i_{c_2} = 60-25 \sin \omega t+0{\cdot}515+0{\cdot}09 \sin^2 \omega t-0{\cdot}43 \sin \omega t$

Therefore, subtracting the expression for i_{c_2} from that for i_{c_1} we have that

$$v_{\text{out}} = K[15{\cdot}11+55{\cdot}9 \sin \omega t+(0{\cdot}01/2)\,(1-\cos 2\,\omega t)] \text{ volts.}$$

The ratio of second harmonic to fundamental frequency amplitude in the output is clearly

$$\frac{0{\cdot}01}{2\times 55{\cdot}9} = \frac{1}{111{\cdot}8}\,\% = 0{\cdot}009\,\%$$

(that is, slightly less than 0·01 %).

This answer shows that, although the two transistors have gain characteristics differing by about 17 %, the effect of using these transistors in Class A push–pull is to give a distortion of less than 0·01 %. The result assumes ideal input and output transformers.

A.7.7

State the meaning of Class A, B and C as applied to power amplifiers. What is meant by "angle of flow"?

A Class C transistor amplifier operates at 10 Mc/s with a collector load tuning capacitor of 1000 pF and a resultant Q-factor of 12·6. The d.c. supply is 40 V, the collector dissipation is 1 W and the collector efficiency is 75 %.

Calculate approximate values for the following:

Effective dynamic resistance of the tuned load circuit.
Power fed to the load.
Direct current drawn from the supply.
Peak-to-peak value of the output voltage.

[H.N.D. 3, 1963]

Q.7.7

Power amplifiers employing transistors may be classified into three groups, from the point of view of a cycle of input current:

Class A: collector current flows during the whole of the input cycle.

Class B: collector current flows during half the input cycle.

Class C: collector current flows for less than half of the input cycle.

The "angle of flow" in an amplifier is proportional to the fraction of the input cycle for which output current flows from one transistor in the amplifier. For example, in a Class C amplifier the angle of flow is less than 180 degrees.

The dynamic resistance of the tuned circuit of the 10 Mc/s amplifier is given by:

$$R_D = Q\omega_0 L = \frac{Q}{\omega_0 C}\ \Omega,$$

where $C = 1000$ pF and $\omega_0 = 20\pi \times 10^6$.

Therefore $$R_D = \frac{12 \cdot 6}{20\pi . 10^6 \times 10^3 \times 10^{-12}} = 200\ \Omega.$$

If the collector dissipation is 1 W and the efficiency is 75%, then $P_{out} = \underline{3\text{W}}$, to the load.

The total input power to the amplifier is, therefore, 4 W, and hence the direct current drawn is 100 Am from a 40 V supply.

The r.m.s. output voltage is $v_{out} = \sqrt{(R_D P_{out})}$ or $\sqrt{(200 \times 3)}$ $= 24 \cdot 5$ V.

Therefore the peak-to-peak output voltage is

$$2\sqrt{2}(v_{out}) = \underline{71\text{ V.}}$$

Q.7.8

A transistor in common emitter connection has the characteristics given in Table 7.3. It is used as a power amplifier in the circuit of Fig. 7.8.1.

TABLE 7.3

Collector voltage (volts)	Collector current I_c (amps)				
	I_b=0 mA	I_b=40 mA	I_b=80 mA	I_b = 120 mA	I_b = 160 mA
0	0	—	—	—	—
−0·5	—	2·0	3·42	4·71	5·9
−5	—	—	—	4·75	5·95
−10	—	2·04	3·50	4·79	6·0
−20	0·008	2·08	3·58	—	—
−30	—	2·12	—	—	—
−40	0·016	—	—	—	—

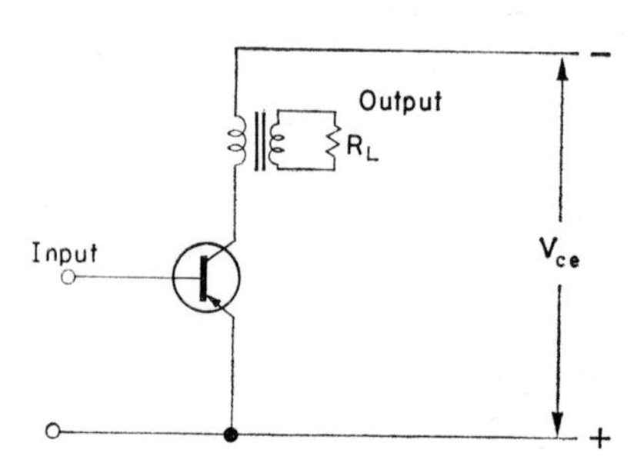

FIG. 7.8.1. Class A power amplifier circuit.

The maximum instantaneous ratings of the transistor are: collector voltage not to exceed −35 V; collector current not to exceed 6 A. Using reasonable approximations choose a suitable value of load resistance R_L and collector supply voltage V_{CC} to give the maximum possible power output. Determine the peak-to-peak base current drive required and the output power developed. Determine also the magnitude and the output power developed. Determine also the magnitude of the fundamental and of the second harmonic component of the output current. Neglect transformer imperfections.

[B.Sc. 3, 1963]

A.7.8

If the maximum instantaneous collector voltage rating of the transistor is -35 V, the voltage across the transistor should be limited to about 30 V in practice.† A load line corresponding

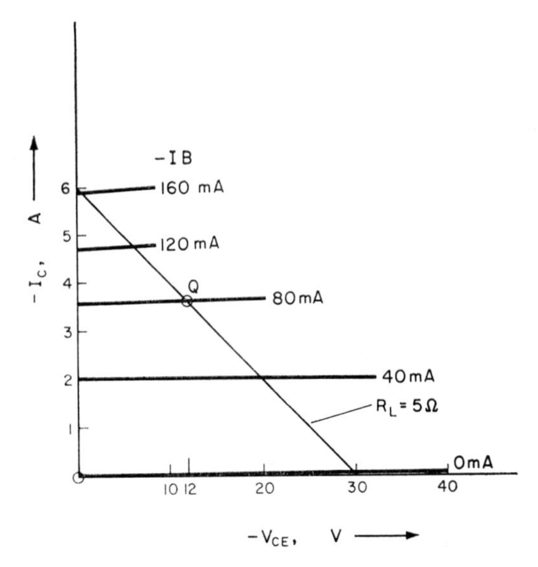

FIG. 7.8.2. Output characteristics, with 5 Ω load line.

to this voltage and the maximum current rating of 6 A is drawn across the output characteristics which are in Fig. 7.8.2.

To allow for the maximum suitable swing of base current the operating point Q has been chosen at the intersection of the load line (of 5 Ω) and the characteristic for $I_B = 80$ mA, which thus requires a supply voltage of $V_{CC} = -12$ V.

The required peak-to-peak base current drive is therefore 160 mA to give maximum power output. The peak-to-peak

† The reason for using a maximum voltage slightly less than the maximum voltage rating is covered in the companion volume, *Semiconductors: Theory, Design and Experiment*, Section 7.2.

values of collector current and voltage corresponding to this input are:

$$\Delta I_c = 5{\cdot}9 - 0{\cdot}012 = 5{\cdot}9 \text{ A}$$

and $$\Delta V_{CE} = 30 - 0{\cdot}5 = 29{\cdot}5 \text{ V}.$$

Hence the available output power is:

$$P_{\text{out}} = \left(\frac{5{\cdot}9}{2\sqrt{2}}\right)\left(\frac{29{\cdot}5}{2\sqrt{2}}\right) = 21{\cdot}7 \text{ W} \quad \text{(say 21 W).}$$

With the maximum drive as specified above the collector current swings from 3·5 A to 5·9 A and to 0 A. This unsymmetrical current waveform indicates the presence of a second harmonic component. The peak-to-peak value of this second harmonic current is $3{\cdot}5 - 2{\cdot}4 = 1{\cdot}1$ A and the peak-to-peak current at the fundamental frequency is 5·9 A. Hence the second harmonic percentage is $(1{\cdot}1/5{\cdot}9) \times 100 = 17{\cdot}5\%$. The reason for taking the sum and difference of the maximum values to determine the relative amounts of fundamental and second harmonic has already been explained in A.7.4.

Q.7.9

Define the following terms in relation to a tuned collector r.f. amplifier: Class C operation, collector efficiency, collector dissipation, angle of flow.

Such an amplifier operated at $4 \times 10^7/2\pi$ c/s.

The supply voltage is 40 V and the collector dissipation is 0·5 W with a collector efficiency of 80%. The effective tuning capacitance is 500 pF and the operating Q is 5.

Calculate the a.c. power fed to the load, the mean current taken by the stage and the peak-to-peak value of the output voltage.

[H.N.C. 2, 1963]

A.7.9

Class C operation and angle of flow are defined in A.7.7, which also involves a very similar calculation.

The collector efficiency of a transistor amplifier is given by

$$\eta = \frac{\text{a.c. output power}}{\text{available collector power}} \times 100\%.$$

In the case of an inductively loaded amplifier the available collector power may be taken as the d.c. input power.

The collector dissipation is taken to be the difference between the available collector power and the output power. Again, with an inductive load, collector dissipation = d.c. input power − a.c. output power.

In this case, power fed to the load $= 0{\cdot}5 \times 4 = 2$ W.

$$\text{Hence mean current from the supply} = \frac{2{\cdot}5\ \text{W}}{40\ \text{V}} = 63\ \text{mA}.$$

Dynamic resistance of the load

$$= \frac{5}{4 \times 10^7 \times 500 \times 10^{-12}} = 250\ \Omega.$$

Peak-to-peak output voltage $= (2\sqrt{2})\sqrt{[(250)2]}$,

i.e. V_{out} (peak-to-peak) $= \underline{63\ \text{V}}$.

CHAPTER 8

Oscillator Circuits

Q.8.1

Sketch the circuit of any type of LC transistor oscillator. Assuming low frequency sinusoidal operation, derive an expression for the frequency of operation and the condition for oscillations to be initiated. Use the low frequency equivalent T circuit for your analysis and state any approximations made.

Why does the above analysis not apply to the normal operating conditons of such an oscillator?

[H.N.C. 3, 1964]

A.8.1

Figure 8.1.1 show the circuit of a tuned collector oscillator, while Fig 8.1.2 gives the corresponding equivalent T circuit. In the latter the following assumptions have been made:

(a) The collector resistance is high enough to be ignored.
(b) The bias components R_B and C_B have negligible effect on the a.c. equivalent circuit.

(c) The resistance of the coil L_2 is negligible,

Applying Kirchhoff's laws to the equivalent circuit,

$$i_e(r_e+r_b+j\omega L_2)-\alpha_0 i_e(r_b+j\omega L_2)\pm j\omega Mi = 0, \quad (1)$$

$$i(r+j\omega L_1+1/j\omega C)-\alpha_0 i_e \cdot (1/j\omega C)\pm j\omega Mi_e(1-\alpha_0) = 0, \quad (2)$$

putting $R = r_e+r_b(1-\alpha_0)$ eqn. (1) reduces to

$$i_e[R+j\omega L_2(1-\alpha_0] = \mp j\omega Mi.$$

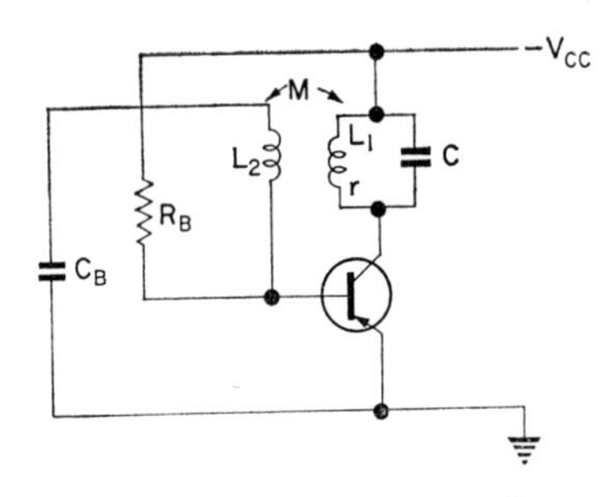

FIG. 8.1.1. Tuned collector oscillator.

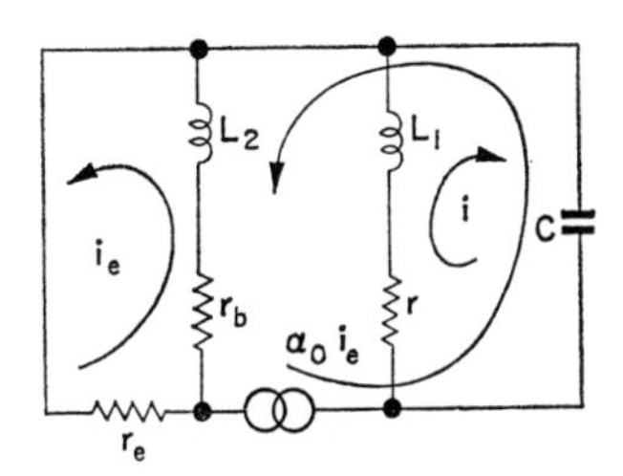

FIG. 8.1.2. Equivalent T circuit of tuned collector oscillator.

From (2)

$$i_e\left[\frac{\alpha_0}{j\omega C}\mp j\omega M(1-\alpha_0)\right] = i\left(r+j\omega L_1+\frac{1}{j\omega C}\right).$$

Hence dividing and cross-multiplying

$$[R+j\omega L_2(1-\alpha_0)]\left[r+j\omega L_1+\frac{1}{j\omega C}\right]$$
$$= \mp j\omega M\left[\frac{\alpha_0}{j\omega C}\mp j\omega M(1-\alpha_0)\right].$$

Equating imaginary parts

$$R\left(\omega L_1-\frac{1}{\omega C}\right)+\omega L_2(1-\alpha_0)r = 0,$$

i.e.

$$\omega^2 L_0 CR-R+\omega^2 L_2 C(1-\alpha_0)r = 0$$

or

$$\omega^2 = \frac{R}{L_1 CR+L_2 C(1-\alpha_0)r}.$$

Therefore

$$f = 1\Big/ 2\pi\sqrt{\left\{L_1 C\left[1+\frac{L_2}{L_1}\times\frac{r}{R}(1-\alpha_0)\right]\right\}}.$$

This equation gives the frequency of oscillation.

Equating real parts,

$$Rr-\omega^2 L_1 L_2(1-\alpha_0)+\frac{L_2}{C}(1-\alpha_0) = \pm\frac{\alpha_0 M}{C}-\omega^2 M^2(1-\alpha_0),$$

i.e. $$\mp M = \frac{CRr}{\alpha_0}+L_2\left(\frac{1-\alpha_0}{\alpha_0}\right)-\omega^2(1-\alpha_0)(L_1 L_2-M^2)$$

If we assume tight coupling such that $M^2 = L_1 L_2$, and take the positive value of M, this equation gives

$$M = \frac{CRr}{\alpha_0}+\frac{L_2}{\beta_0}, \qquad \text{where} \qquad \beta_0 = \frac{\alpha_0}{1-\alpha_0}.$$

This expression enables the required value of the mutual inductance M, and hence L_1, to be determined.

In practice transistor oscillators of this type usually operate under Class AB, Class B or Class C conditions with current only flowing for part of the cycle. Thus, although the above analysis is true for the initial build up of oscillations, it will not be valid once stable conditions are attained. The effects of loading have not been considered, but if this is small it may be allowed for by an increase in the series resistance r.

For further details see *T.D.E.*, sections 8.1.1 and 8.1.2.

Q.8.2

Show that if sinusoidal operation is assumed the Colpitts type oscillator (Fig. 8.2.1) oscillates at a frequency given by

$$\omega^2 = \frac{1}{LC}\left[1+\frac{r(1-\alpha_0)}{R}\cdot\frac{C}{C_1}\right], \quad \text{where} \quad \frac{1}{C} = \frac{1}{C_1}+\frac{1}{C_2}$$

and $R = r_e+r_b(1-\alpha_0)$.

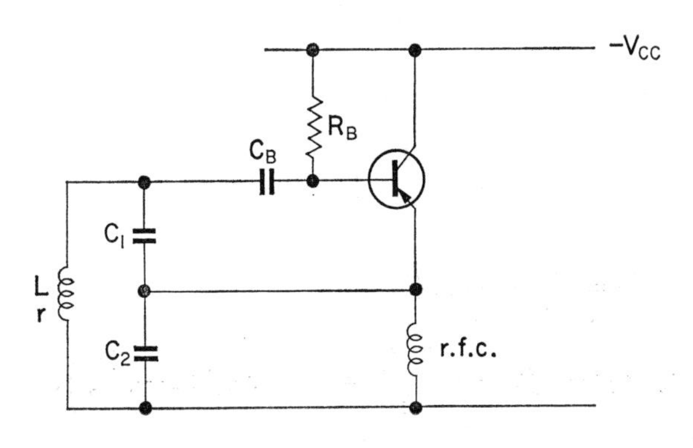

FIG. 8.2.1. Colpitts oscillator.

Determine also the condition for oscillations to be initiated. The effect of the bias circuit and r.f. choke may be neglected.

A.8.2

Figure 8.2.2 shows the T equivalent circuit of the oscillator and as in the previous question the value of the collector resistance r_c is assumed high enough to be ignored.

Applying Kirchhoff's laws,

$$i_e\left(r_e+r_b+\frac{1}{j\omega C_1}\right)+\frac{i}{j\omega C_1}-\alpha_0 i_e r_b = 0, \tag{1}$$

$$i\left(r+j\omega L+\frac{1}{j\omega C_1}+\frac{1}{j\omega C_2}\right)+\frac{i_e}{j\omega C_1}+\alpha_0 i_e(r+j\omega L) = 0. \tag{2}$$

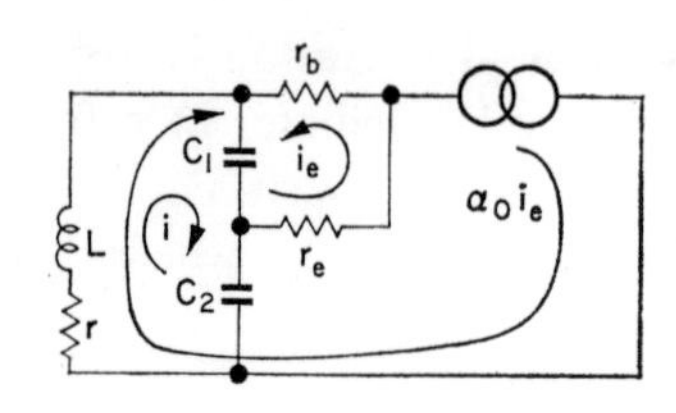

FIG. 8.2.2. Equivalent T circuit of Colpitts oscillator.

Putting $R = r_e+r_b(1-\alpha_0)$ in (1),

$$i_e\left(R+\frac{1}{j\omega C_1}\right) = -\frac{i}{j\omega C_1}.$$

Putting $\frac{1}{C} = \frac{1}{C_1}+\frac{1}{C_2}$ in (2),

$$i_e\left(\alpha_0 r+\alpha_0 j\omega L+\frac{1}{j\omega C_1}\right) = -i\left(r+j\omega L+\frac{1}{j\omega C}\right).$$

Hence dividing and cross-multiplying,

$$\left(R+\frac{1}{j\omega C_1}\right)\left(r+j\omega L+\frac{1}{j\omega C}\right) = \frac{1}{j\omega C_1}\left(\alpha_0 r+\alpha_0 j\omega L+\frac{1}{j\omega C_1}\right).$$

Equating imaginary parts

$$R\omega L-\frac{R}{\omega C}-\frac{r}{\omega C_1} = -\frac{\alpha_0 r}{\omega C_1}$$

or

$$\omega^2 LCC_1R = C_1R+Cr(1-\alpha_0),$$

i.e. $$\omega^2 = \frac{1}{LC}\left[1+\frac{r(1-\alpha_0)}{R}\times\frac{C}{C_1}\right]$$

or $$f = \frac{1}{2\pi\sqrt{LC}}\times\sqrt{\left[1+\frac{r(1-\alpha_0)}{R}\times\frac{C}{C_1}\right]}.$$

Equating real parts,

$$Rr+\frac{L}{C_1}-\frac{1}{\omega^2CC_1} = \frac{\alpha_0 L}{C_1}-\frac{1}{\omega^2C_1^2}.$$

But $\omega^2 = \dfrac{1}{LC}$,

$$Rr = \frac{\alpha_0 L}{C_1}-\frac{LC}{C_1^2},$$

i.e. $$\alpha_0 = \frac{C}{C_1}+\frac{C_1Rr}{L}.$$

This expression gives the condition for oscillations to be initiated.

Q.8.3

Show that the circuit given in Fig. 8.3.1 will oscillate if

$$M = \frac{RrC}{\alpha_0}+\frac{L_1}{\alpha_0},$$

where $$R = r_e+r_b\,(1-\alpha_0).$$

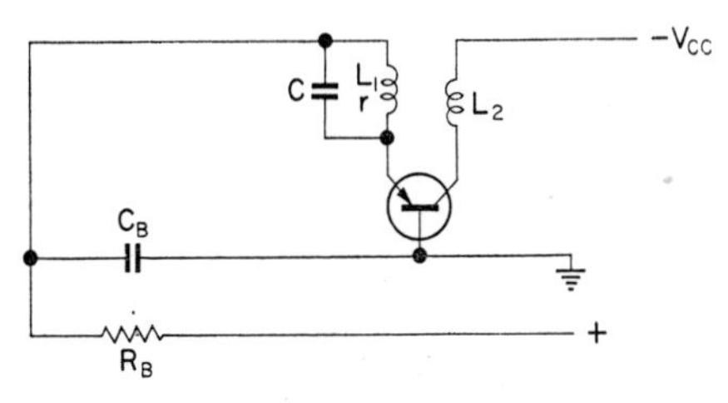

FIG. 8.3.1. Tuned emitter oscillator.

Determine the frequency of oscillation of the circuit.

Assume sinusoidal operation and ignore the effect of the bias circuit.

A.8.3

Fig. 8.3.2 represents the equivalent *T* circuit for the oscillator. The collector resistance is assumed to be very high and the resistance of coil L_2 very low.

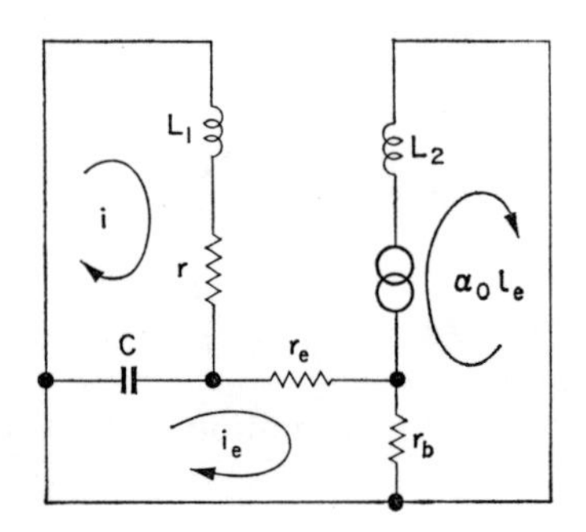

FIG. 8.3.2. Equivalent *T* circuit of tuned emitter oscillator.

The circuit equations are

$$i_e\left(r_e+r_b+\frac{1}{j\omega C}\right)-\frac{1}{j\omega C}-\alpha_0 i_e r_b = 0, \qquad (1)$$

$$i\left(r+j\omega L_1+\frac{1}{j\omega C}\right)-\frac{i_e}{j\omega C}\pm j\omega M\alpha_0 i_e = 0. \qquad (2)$$

Putting $R = r_e+r_b(1-\alpha_0)$ in (1),

$$i_e\left(R+\frac{1}{j\omega C}\right) = \frac{i}{j\omega C}.$$

From (2)

$$i_e\left(\frac{1}{j\omega C}\mp j\omega M\alpha_0\right) = i\left(r+j\omega L_1+\frac{1}{j\omega C}\right).$$

Hence diving and cross-multiplying,

$$\left(R+\frac{1}{j\omega C}\right)\left(r+j\omega L_1+\frac{1}{j\omega C}\right) = \frac{1}{j\omega C}\left(\frac{1}{j\omega C}\mp j\omega M\alpha_0\right).$$

Equating real parts,

$$Rr+\frac{L_1}{C}-\frac{1}{\omega^2 C^2} = -\frac{1}{\omega^2 C}\mp\frac{M\alpha_0}{C},$$

i.e.

$$\mp M = \frac{CRr}{\alpha_0}+\frac{L_1}{\alpha_0}.$$

The positive value of M gives the condition for oscillations ot be initiated, i.e.

$$M = \frac{CRr}{\alpha_0}+\frac{L_1}{\alpha_0}.$$

Equating imaginary parts,

$$-\frac{r}{\omega C}+\omega L_1 R-\frac{R}{\omega C} = 0,$$

i.e.

$$\omega^2 L_1 CR = R+r,$$

i.e.

$$\omega^2 = \frac{1}{L_1 C}\left(1+\frac{r}{R}\right)$$

or

$$f \equiv \frac{1}{2\pi\sqrt{(L_1 C)}}\sqrt{\left(1+\frac{r}{R}\right)}.$$

Q.8.4

A transistor amplifier has a very low input impedance, a very high output impedance and a current gain of up to 100 with no phase reversal. It is connected as shown in Fig. 8.4.1 with an RC feedback network.

Determine whether the circuit is capable of oscillating; if so, at what frequency and with what minimum value of current gain in the amplifier?

[H.N.C. 2, 1960]

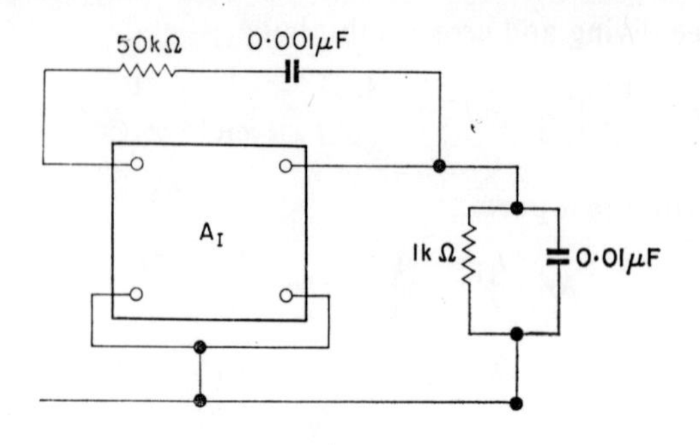

FIG. 8.4.1. Two-stage RC oscillator.

A.8.4

The circuit corresponds to a constant current source feeding a RC network as shown in Fig. 8.4.2.

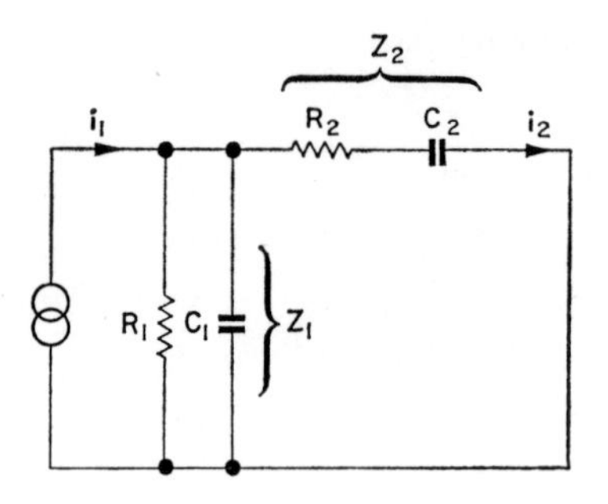

FIG. 8.4.2. Feedback network for two-stage RC oscillator.

For the series arm,
$$Z_2 = R_2 + \frac{1}{j\omega C_2}$$
$$= \frac{1+j\omega C_2 R_2}{j\omega C_2}.$$

For the shunt arm,
$$Z_1 = \frac{R_1/j\omega_1}{R_1+1/j\omega_1}$$
$$= \frac{R_1}{1+j\omega C_1 R_1}.$$

But $(i_1 - i_2) Z_1 = i_2 Z_2$,

therefore
$$\frac{i_1}{i_2} = \frac{Z_1 + Z_2}{Z_1} = 1 + \frac{Z_2}{Z_1}$$
$$= 1 + \frac{(1 + j\omega C_2 R_2)}{j\omega C_2} \times \frac{(1 + j\omega C_1 R_1)}{R_1}$$
$$= 1 + \frac{1}{j\omega C_2 R_1}(1 - \omega^2 C_1 C_2 R_1 R_2 + j\omega C_1 R_1 + j\omega C_2 R_2)$$
$$= \left(1 + \frac{C_1}{C_2} + \frac{R_2}{R_1}\right) - j\left(\frac{1 - \omega^2 C_1 C_2 R_1 R_2}{\omega C_2 R_1}\right).$$

This equation relates the output current of the amplifier to the feedback current.

They are in phase when
$$1 - \omega^2 C_1 C_2 R_1 R_2 = 0,$$

i.e.
$$\omega^2 = \frac{1}{C_1 C_2 R_1 R_2}.$$

Then
$$\frac{i_1}{i_2} = 1 + \frac{C_1}{C_2} + \frac{R_1}{R_2}.$$

The circuit will oscillate if the current gain of the amplifier is greater than this ratio.

With the values given,
$$\frac{i_1}{i_2} = 1 + \frac{0{\cdot}01}{0{\cdot}001} + \frac{50}{1}$$
$$= \underline{61}.$$

This is the minimum value of current gain required for the circuit to oscillate. Since the current gain is up to a 100, oscillations may be initiated.

The frequency of oscillation is given by
$$\omega^2 = \frac{1}{0{\cdot}01 \times 10^6 \times 0{\cdot}001 \times 10^{-6} \times 10^3 \times 50 \times 10^3}$$
$$= 0{\cdot}2 \times 10^{10},$$

i.e.
$$f = \underline{71{\cdot}2 \text{ kc/s}}.$$

Q.8.5

Derive an expression for the approximate frequency of oscillation for the RC oscillator shown in Fig. 8.5.1. Assume the load and bias resistors R_L and R_B are very high, the input

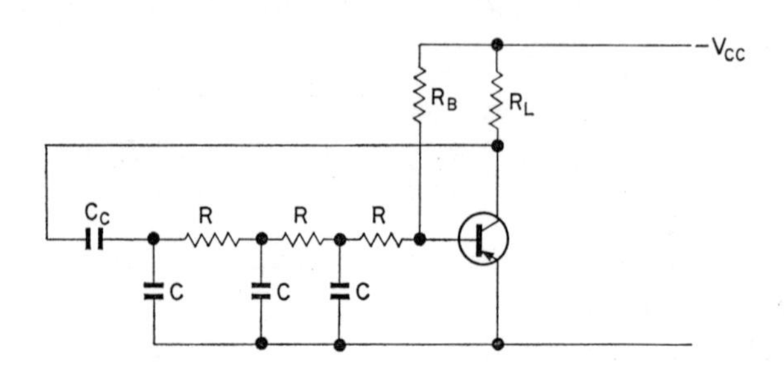

FIG. 8.5.1. Single-stage RC oscillator.

resistance is low and the reactance of the coupling capacity C_C may be ignored.

Evaluate when $R = 2{\cdot}2\ \text{k}\Omega$, $C = 0{\cdot}01\ \mu\text{F}$.

A.8.5

In the RC oscillator shown the phase shift produced by the transistor is counteracted by the phase shift in the RC network.

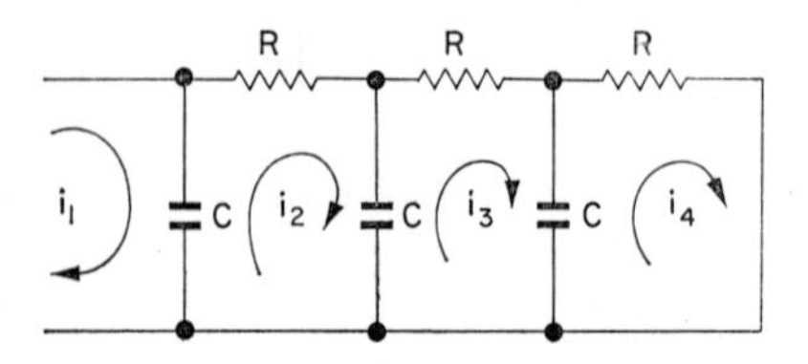

FIG. 8.5.2. Feedback network for single-stage RC oscillator.

Assuming the transistor is operating well below its cut-off frequency the transistor itself produces a phase shift of 180 degrees and hence for oscillation the output current of the network shown in Fig. 8.5.2 must be in antiphase to the input current.

The circuit equations are:

$$i_2\left(R+\frac{2}{j\omega C}\right)-\frac{i_1}{j\omega C}-\frac{i_3}{j\omega C} = 0, \qquad (1)$$

$$i_3\left(R+\frac{2}{j\omega C}\right)-\frac{i_2}{j\omega C}-\frac{i_4}{j\omega C} = 0, \qquad (2)$$

$$i_4\left(R+\frac{1}{j\omega C}\right)-\frac{i_3}{j\omega C} = 0. \qquad (3)$$

From (3)

$$i_3 = i_4(1+j\omega CR).$$

Hence from (2)

$$i_4(1+j\omega CR)\left(R+\frac{2}{j\omega C}\right)-\frac{i_2}{j\omega C}-\frac{i_4}{j\omega C} = 0$$

or

$$i_2 = i_4[(1+j\omega CR)(2+j\omega CR)-1]$$

$$= i_4[(1-\omega^2C^2R^2)+3j\omega CR].$$

Substituting for i_2 and i_3 in (1),

$$i_4[(1-\omega^2C^2R^2)+3j\omega CR]\left(R+\frac{2}{j\omega C}\right)-\frac{i_1}{j\omega C}-i_4\,\frac{(1+j\omega CR)}{j\omega C}=0,$$

i.e. $$i_4\{[(1-\omega^2C^2R^2)+3j\omega CR](2+j\omega CR)-(1-j\omega CR)\} = i_1,$$

therefore $$i_4 = \frac{1}{(1-5\omega^2C^2R^2)+j\omega C(6-\omega^2C^2R^2)}.$$

For no j term

$$6-\omega^2C^2R^2 = 0,$$

i.e. $$\omega = \frac{\sqrt{6}}{CR}.$$

At this frequency

$$\frac{i_4}{i_1} = -\frac{1}{29}.$$

Hence the circuit will oscillate at a frequency given by $f = \sqrt{6}/2\pi CR$ if the current gain of the amplifier is -29.

With the values given

$$f = \frac{\sqrt{6}}{2\pi \times 0{\cdot}01 \times 10^{-6} \times 2{\cdot}2 \times 10^{3}}$$

$$= \underline{17{\cdot}7 \text{ kc/s.}}$$

Q.8.6

Show that the current gain of a grounded base transistor is approximately equal to the current amplification factor α if the impedance connected between the collector and the base is much less than the collector resistance r_c.

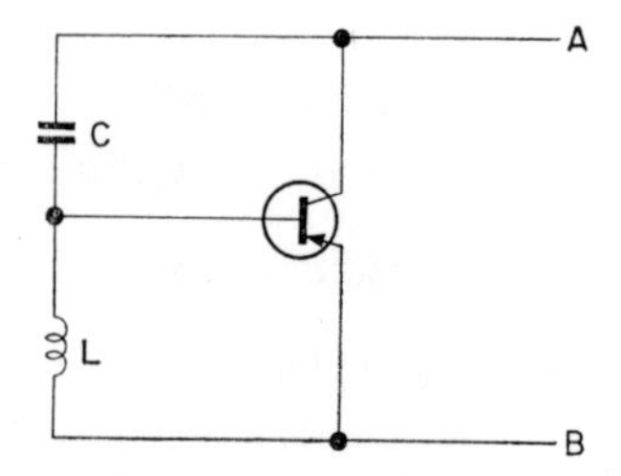

FIG. 8.6.1. Negative resistance transistor circuit.

In the circuit of Fig. 8.6.1 the reactance of the capacitor $-X_C$ is much less than r_c at the frequencies of interest. If X_L is the reactance of the inductor, show that the condition $X_L/X_c < \alpha$ must be aproximately satisfied if the input impedance at the terminals AB has a negative resistive part. Hence, determine the nature and magnitude of the reactance, X, in Fig. 8.6.2 to cause oscillation at a frequency for which the reactances of the inductor and the capacitor are 500 Ω and $-625\ \Omega$ respectively. What is the maximum value of R for which the circuit will oscillate? Any reasonable approximations may be used.

The parameters of the transistor in Fig. 8.6.2 are:

$$r_e = 59\ \Omega, \qquad r_b = 500\ \Omega; \qquad r_c = 500\ \mathrm{K}\Omega; \qquad \alpha = 0{\cdot}98$$

[B.Sc. 3., 1961]

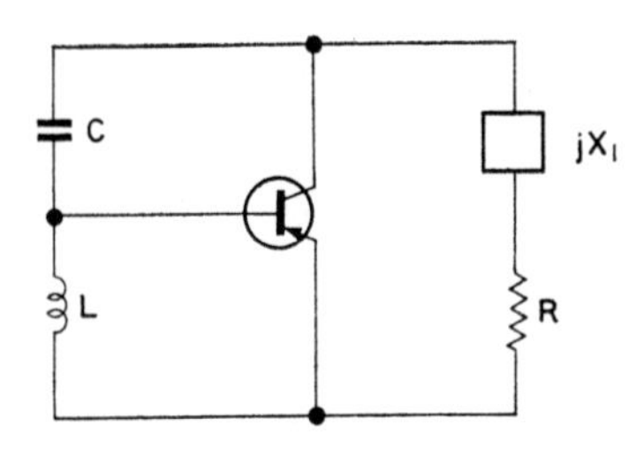

FIG. 8.6.2. Negative resistance transistor oscillator.

A.8.6

The equivalent circuit of a grounded base transistor with an impedance Z connected between collector and base is shown in Fig. 8.6.3. Applying Kirchhoff's laws to the output circuit,

$$i_c(r_b+r_c+Z)+i_e r_b+\alpha i_e r_c = 0.$$

Therefore
$$\frac{i_c}{i_e} = -\frac{r_b+\alpha r_c}{r_b+r_c+Z}.$$

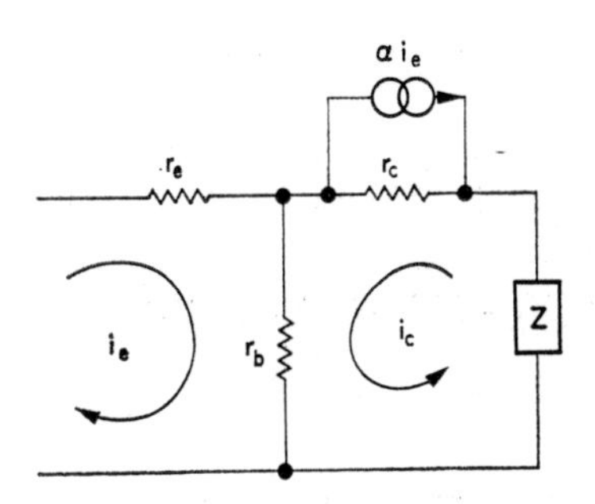

FIG. 8.6.3. Grounded base equivalent circuit.

In practice $\alpha r_c \gg r_b$,

therefore
$$\frac{i_c}{i_e} = -\frac{\alpha r_c}{r_c+Z}$$
$$= -\frac{\alpha}{1+Z/r_c}.$$

Hence, if $Z \ll r_c$, the current gain $i_c/i_e = -\alpha$.

The above approximation may be used to find the impedance at the terminals AB. The equivalent circuit is shown in Fig. 8.6.4 and, applying a voltage v to the terminals AB,

$$v = (i+\alpha i_e)(-jX_C)+(i+i_e)jX_L, \tag{1}$$
$$0 = i_e r_e+(1-\alpha)i_e r_b+(i+i_e)jX_L. \tag{2}$$

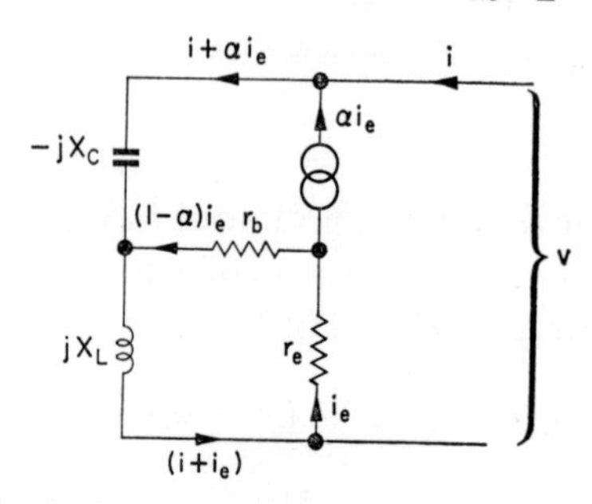

FIG. 8.6.4. Equivalent circuit of negative resistance transistor oscillator.

From (2)
$$i_e = \frac{-jX_L}{r_e+r_b(1-\alpha)+jX_L}\cdot i$$
$$= -\frac{jX_L}{R+jX_L}i, \quad \text{where} \quad R = r_e+r_br(1-\alpha).$$

Hence from (1)
$$v = ji(X_L-X_C)+ji_e(X_L+\alpha X_C)$$
$$= ji(X_L-X_C)+\frac{X_L i}{R+jX_L}(X_L+\alpha X_C)$$
$$= ji(X_L-X_C)+\frac{X_L i}{R^2+X_L^2}(X_L-\alpha X_C)(R-jX_L).$$

Therefore

$$Z_{AB} = \frac{v}{i} = \left[\frac{RX_L}{R^2+X_L^2}(X_L-\alpha X_C)\right]$$
$$+j\left[X_L-X_C-\frac{X_L^2(X_L-\alpha X_C)}{R^2+X_L^2}\right].$$

For this impedance to have a negative resistance component $X_L-\alpha X_C$ must be negative, i.e.

$$X_L < \alpha X_C \quad \text{or} \quad \frac{X_L}{X_C} < \alpha.$$

With the values given, $R = 50+500(1-0{\cdot}98) = 60\ \Omega$. Therefore

$$Z_{AB} = \left[\frac{60\times 500(500-0{\cdot}98\times 625)}{60^2+500^2}\right]$$
$$+j\left[\frac{500-625-500^2(500-0{\cdot}98\times 625)}{60^2+500^2}\right]$$
$$= \underline{-13{\cdot}3-j14{\cdot}1\ \Omega}$$

Hence the circuit shown in Fig. 8.6.2 will oscillate if the reactance X is inductive and $14{\cdot}1\ \Omega$ in magnitude.

The resistance R will damp the oscillations and its maximum value for which the circuit will oscillate is $13.3\ \Omega$.

Q.8.7

Sketch the characteristics of an Esaki or tunnel diode indicating approximate scales of voltage and current. Explain briefly its importance as a circuit element.

Figure 8.7.1 indicates a possible equivalent circuit for the tunnel diode. Derive an expression for the input impedance and determine the frequencies for which the real and quadrature components become zero if $R = 1{\cdot}5\ \Omega$, $L = 0{\cdot}01\ \mu$H, $C = 50$ pF, $r = -25\ \Omega$. What is the significance of the results?

[I.E.E., June 1963]

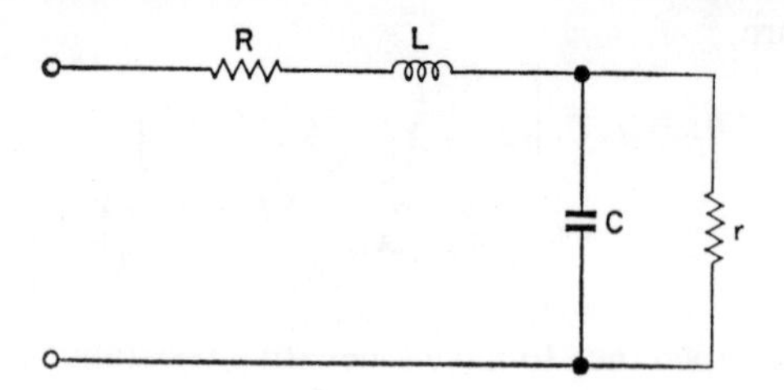

FIG. 8.7.1. Equivalent circuit of a tunnel diode.

A.8.7

A typical characteristic for a tunnel diode is shown in Fig. 8.7.2. Between C and A and beyond B an increase in voltage leads to an increase in current, but over the range AB the reverse occurs giving rise to a negative resistance.

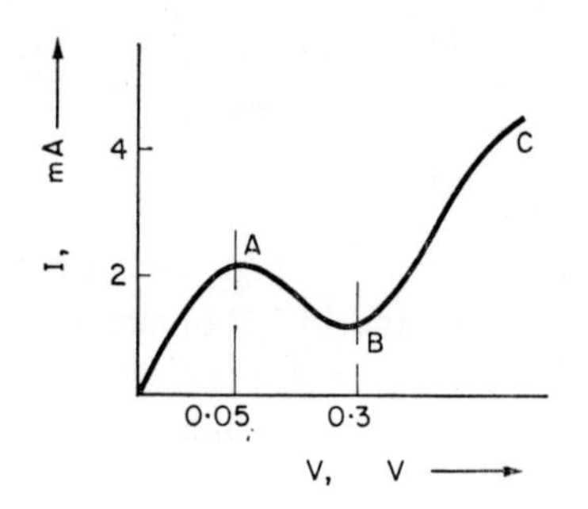

FIG. 8.7.2. Characteristic of a tunnel diode.

The importance of tunnel diodes as circuit elements may be summarized under the following headings.

1. *Physical operation.* The "tunnelling" action of the electron carriers in such diodes takes place at approximately the speed of light. This enables the diodes to be used as circuit elements at microwave frequencies.

2. *Size.* The small physical size of such diodes enables stray capacities and lead inductances to be kept to a minimum. This

is very useful in the design and construction of circuits at v.h.f.

3. *Circuitry.* The fact that the diode is an active two terminal device enables the circuits used to be relatively simple. Difficulties arise with amplifier circuits since input and output circuits cannot be isolated.

4. *Applications.* The regenerative effect of the negative resistance leads to the use of tunnel diodes in amplifiers, oscillators and switching circuits over a wide range of frequency.

The input impedance of the circuit shown in Fig. 8.7.1 is given by

$$Z_{\text{in}} = R + j\omega L + \frac{r/j\omega C}{r + 1/j\omega C}$$

$$= R + j\omega L + \frac{r}{1 + j\omega Cr}$$

$$= R + j\omega L + \frac{r(1 - j\omega Cr)}{1 + \omega^2 C^2 r^2}$$

$$= \left(R + \frac{r}{1 + \omega^2 C^2 r^2}\right) + j\omega\left(L - \frac{Cr^2}{1 + \omega^2 C^2 r^2}\right).$$

The real term is zero if

$$R + \frac{r}{1 + \omega^2 C^2 r^2} = 0,$$

i.e.

$$1 + \omega^2 C^2 r^2 + \frac{r}{R} = 0,$$

$$\omega = \frac{1}{Cr}\sqrt{\left[-\left(1 + \frac{r}{R}\right)\right]}.$$

Since r is negative,

$$f = \frac{1}{2\pi Cr}\sqrt{\left(\frac{|r|}{R} - 1\right)}.$$

The quadrature term is zero if

$$L - \frac{Cr^2}{1+\omega^2 C^2 r^2} = 0,$$

$$1+\omega^2 C^2 r^2 - \frac{Cr^2}{L} = 0,$$

$$\omega^2 = \frac{1}{LC} - \frac{1}{Cr^2}$$

$$= \frac{1}{LC}\left(1 - \frac{L}{Cr^2}\right),$$

i.e.
$$f = \frac{1}{2\pi\sqrt{(LC)}}\sqrt{\left(1-\frac{L}{Cr^2}\right)}.$$

With the values given the real part is zero when

$$f = \frac{1}{2\pi \times 50 \times 10^{-12} \times 25}\sqrt{\left(\frac{25}{1 \cdot 5} - 1\right)}$$

$$= 504 \text{ Mc/s}.$$

The quadrature term is zero when

$$f = \frac{1}{2\pi\sqrt{(0 \cdot 01 \times 10^{-6} \times 50 \times 10^{-12})}}\sqrt{\left[1 - \frac{0 \cdot 01 \times 10^{-6}}{50 \times 10^{-12}(25)^2}\right]}$$

$$= \underline{186 \text{ Mc/s}}.$$

The latter frequency of 186 Mc/s, where the impedance is a negative resistance and the reactance is zero corresponds to the natural frequency of oscillation of the circuit.

At frequencies below 504 Mc/s the input impedance has a negative resistance component and this is the upper frequency limit for which the circuit is capable of oscillating with additional series capacitance.

Q.8.8

Draw the circuit diagram of a crystal controlled transistor oscillator and explain its operation.

List the factors which determine the frequency stability of such an oscillator.

[C. and G. 3, 1961]

A.8.8

Figure 8.8.1 shows the circuit of a crystal controlled tuned collector oscillator. It is very similar to the oscillator analysed

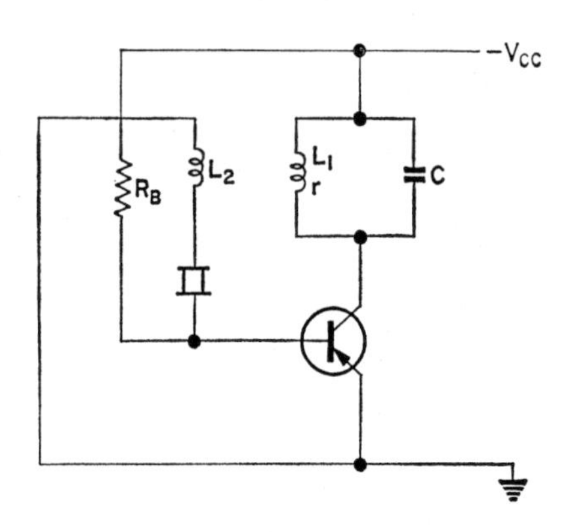

FIG. 8.8.1. Crystal controlled transistor oscillator.

in A.8.1 with feedback from the collector tuned circuit to the base circuit. If the alternating current induced in the base circuit is of correct phase and magnitude, the a.c. power developed at the collector will be sufficient to overcome losses in the tuned circuit and any load that is connected.

The crystal itself behaves as a circuit with two resonant frequencies. The equivalent circuit is shown in Fig. 8.8.2 and the response curve in Fig. 8.8.3. Oscillator circuits may be designed to use the series resonant frequency f_r, the parallel resonant frequency f_a or the narrow frequency band $f_a - f_r$, where the crystal behaves as an inductance. This particular circuit uses the series resonant mode. The circuit is designed to oscillate at a frequency f_r and any tendency for the frequency to drift will be counteracted by a rapid change in reactance of the crystal.

As shown in A.8.1, the frequency of oscillation is given approximately by

$$\omega^2 = 1/L_1C\left[1+\frac{L_2}{L_1}\times\frac{r}{R}(1-\alpha_0)\right].$$

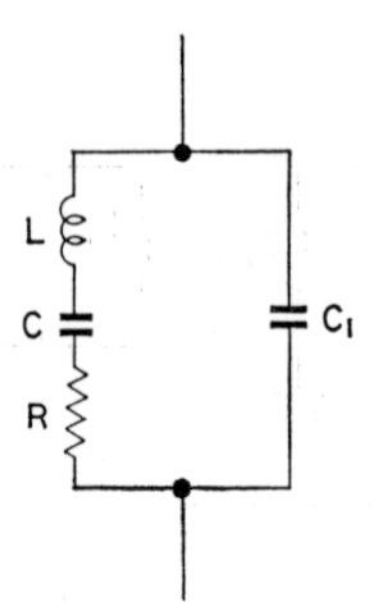

FIG. 8.8.2. Equivalent circuit of a crystal.

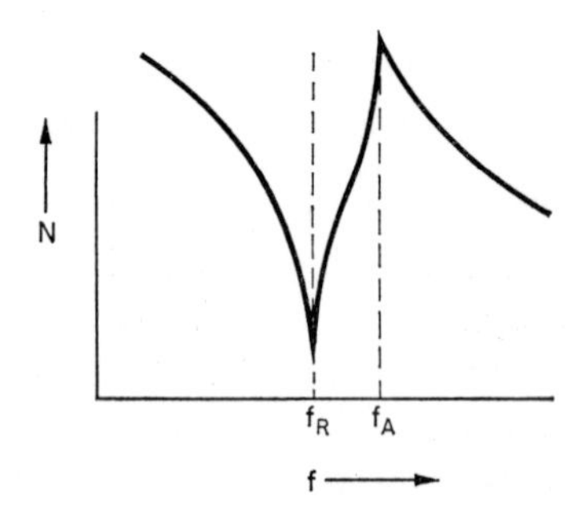

FIG. 8.8.3. Impedance–frequency relationship for a crystal.

Hence, without the crystal the frequency is mainly dependent on the constancy of the tuned circuit components L_1 and C, and to a lesser extent on the transistor parameters themselves.

With a crystal in the feedback path the stability depends mainly on the crystal, particularly the variation of its resonant frequency with temperature. This depends on the crystal "cut", but for highest stability the crystal would be maintained at an even temperature by a crystal oven.

For further details of crystal controlled transistor oscillators see *T.D.E.*, section 8.4.

Q.8.9

List the effects of feedback on the characteristics of an amplifier.

A three-stage transistor amplifier has an overall voltage gain which may be represented by the expression

$$m = \frac{-1000}{(1+j10^{-5}f)^3},$$

where f is the frequency in c/s. A resistive potential divider feeds back a fraction $\beta = 1/150$ of the output voltage in series with the input.

Sketch the polar plot of βm for frequencies between 0 and ∞. Show that the amplifier is stable and calculate the percentage increase in stage gain necessary to induce instability. Mention all approximations and assumptions made.

[I.E.E., Nov. 1962]

A.8.9

Feedback may be positive or negative, the former being rarely used in amplifier circuits. The use of positive feedback is inherent in oscillator design, when instability is required, but early radio receivers also employed positive feedback to increase the selectivity and sensitivity of r.f. amplifiers.

Negative feedback is universally used in high quality amplifiers and may be proportional to either the output voltage or output current. Except in one respect (output impedance) their effects are similar and the two forms will not be considered separately.

The effects of negative feedback are:

(1) The gain is stabilized, against component and supply voltage variations, at the expense of sensitivity.
(2) Better frequency response, i.e. a wider bandwidth.
(3) Less distortion and noise.
(4) Higher input impedance.
(5) Modification of output impedance. This is reduced for voltage negative feedback but increased for current negative feedback.
(6) Phase-frequency response improved.

The feedback factor:

$$\beta m = -\frac{20}{3(1+j10^{-5}f)^3}.$$

Hence Table 8.1 may be derived.

TABLE 8.1

f	$1+j10^{-5}f = R\underline{/\theta}$		βm
0	1	$1\underline{/0}$	$6{\cdot}67\underline{/180°}$
10^2	$1+j10^{-3}$	$1\underline{/0}$	$6{\cdot}67\underline{/180°}$
10^4	$1+j10^{-1}$	$1{\cdot}005\underline{/5\tfrac{1}{2}°}$	$6{\cdot}57\underline{/197°}$
5×10^4	$1+j0{\cdot}5$	$1{\cdot}118\underline{/26\tfrac{1}{2}°}$	$4{\cdot}88\underline{/260°}$
10^5	$1+j1$	$1{\cdot}414\underline{/45°}$	$2{\cdot}36\underline{/315°}$
2×10^5	$1+j2$	$2{\cdot}326\underline{/63\tfrac{1}{2}°}$	$0{\cdot}596\underline{/10°}$
10^6	$1+j10$	$10{\cdot}05\underline{/84\tfrac{1}{2}°}$	$0{\cdot}0065\underline{/73°}$
∞	$1+j\infty$	$\infty\underline{/90}$	$0\underline{/90°}$

The polar plot of βm (or Nyquist diagram) is as shown in Fig. 8.9.1. The diagram does not enclose the point (1·0) and the amplifier is stable.

From the diagram the frequency for a phase shift of 360 degrees is just less than 2×10^5 c/s.

Assuming each stage gives equal gain this occurs when the phase shift per stage is 60 degrees,

i.e. $$\tan 60 = 10^{-5}f$$

or $$f = \underline{173 \text{ kc/s.}}$$

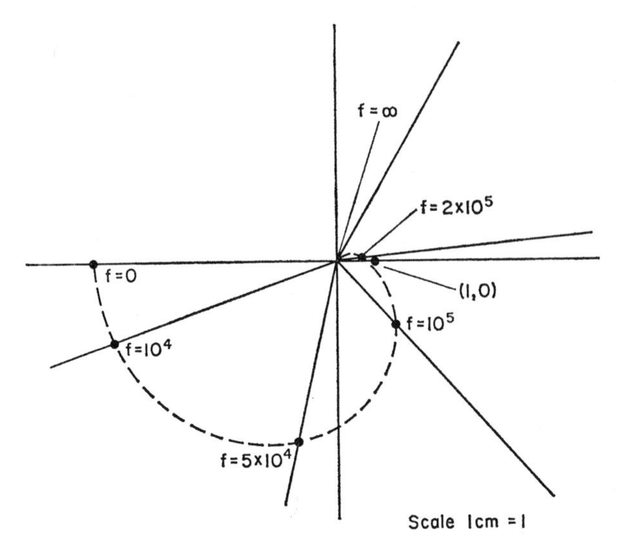

FIG. 8.9.1. Nyquist diagram of a three-stage amplifier.

Then the overall gain without feedback

$$= -\frac{1000}{(1+j1{\cdot}73)^3}$$

$$= 125$$

and $$\beta m = 5/6.$$

Hence instability will be produced if the overall gain is increased by 20%.

If X is the fractional gain per stage,

$$(1+X)^3 = 1{\cdot}2,$$
$$X = 0{\cdot}063,$$

i.e. the increase in stage gain is 6·3%.

Q.8.10

A voltage amplifier has three identical stages, each having a voltage amplification A at frequency f c/s, where

$$A = \frac{50}{1+j(f/f_0)}$$

and $$f_0 = 10^4 \text{ c/s}.$$

Voltage feedback is provided by a resistive network, connected to feed back a fraction β of the output voltage. The feedback is negative at low frequencies. By using the Nyquist diagram, or otherwise, calculate the largest value of β which can be used without the amplifier becoming unstable.

What will be the frequency of the oscillation if the amplifier is unstable and what will determine the amplitude of the oscillation?

[B.Sc. 3, 1962]

A.8.10

This question is very similar to the previous one and it is left to the reader to show that the maximum value of β is 1/15625 and the frequency of oscillation when the amplifier is unstable is 17·3 kc/s.

CHAPTER 9

Pulse and Computing Circuits

Q.9.1

Draw the circuit of a three-input AND gate, using diodes. Discuss in detail the requirements for successful operation of this circuit and point out how this is affected by the non-ideal properties of a semiconductor *pn* junction.

Why is an emitter follower often used between two diode gate circuits in tandem?

[H.N.D. 3, 1963]

A.9.1

The circuit of a three-input AND gate, employing diode-resistor logic, is given as Fig. 9.1.1. Successful operation of this circuit requires that the input is fed from a low resistance source, to complete a d.c. loop, and the output feeds into a circuit of relatively high impedance, which does not draw a significant current from the supply. The diodes must operate as nearly ideal switches over the whole range of frequency for which the gate circuit is intended. The supply voltage V_B must exceed the maximum expected signal voltage at the input.

Frequently, small point-contact diodes are employed in these circuits, since *pn* junction diodes (particularly those made from germanium) have the following disadvantages:

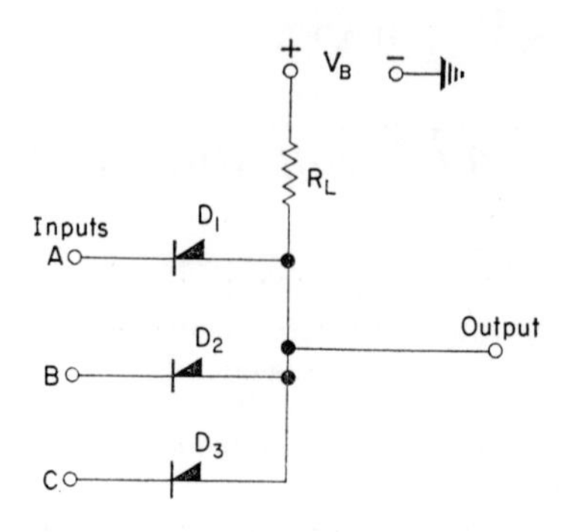

FIG. 9.1.1. Three-input AND gate with diodes.

(a) The diode has finite reverse resistance and some current is thus always being drawn, to impose a load on the input circuit. This effect becomes pronounced with a gate circuit which has many inputs.

(b) Non-zero forward resistance and the forward voltage drop prior to onset of conduction (perhaps 200 mV) combine to give a significant volt drop across each diode. So every stage of diode gating causes a power loss and signal deterioration.

(c) Diode junction capacitance increases the rise and fall times of the voltage pulses. This capacitance is somewhat dependent on d.c. conditions within the diode.

(d) Charge carrier storage, sometimes known as "hole storage" means that the diodes take a finite time to turn ON and OFF. This leads to the appearance of voltage "spikes" on the output waveform.

An emitter follower amplifier is useful between successive diode stages, since it provides power gain and the desirable conditions of high input and low output impedance.

Q.9.2

Explain the significance of the use of binary numbers in electrical systems. Give three examples of decimal numbers and the corresponding binary numbers.

Sketch the diagram of a binary counting (Eccles–Jordan) circuit, using either *pnp* or *npn* transistors, and outline the mode of operation.

[H.N.D. 3, 1963]

A.9.2

The binary system of numbers uses only two digits. Thus a binary number may be represented by the use of a series of bi-stable devices. Many electrical devices are of this form, e.g.: a switch may be open or closed, a relay may be energized or de-energized and a transistor may be conducting or cut-off. To represent a decimal number, however, it is necessary to be able to set a voltage to any one of ten stable values. A signal of this form is easily distorted in transmission and its original value may be lost. On the other hand a two-state signal (e.g.: a voltage either present or absent) is easy to detect and distortion in transmission will not change its state. Thus if numbers are to be handled electrically there is much in favour of the binary system.

Three corresponding decimal and binary numbers are given below

Decimal	Binary
9	1001
27	11011
609	1001100001

A binary counting circuit is shown in Fig. 9.2.1.

The circuit has two stable states, either (a) Q_1 bottomed and Q_2 cut-off or (b) Q_1 cut-off and Q_2 bottomed. It will only change from one state to the other when a trigger pulse is applied.

If Q_1 is cut-off and Q_2 is bottomed then a negative trigger pulse at the base of Q_1 will tend to make Q_1 conduct. This will cause the collector of Q_1 to go positive which in turn makes

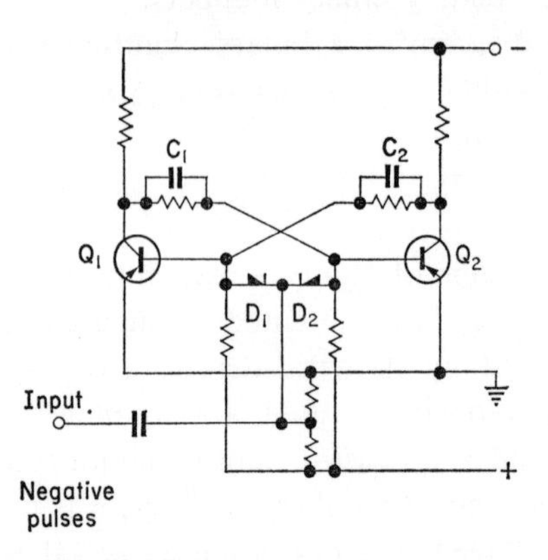

FIG. 9.2.1. Binary counting circuit.

the base of Q_2 go positive. This tends to cut-off Q_2 and causes the collector of Q_2 to go negative and hence the base of Q_1 to go more negative. This process is cumulative and ends with Q_1 conducting and Q_2 cut-off, i.e. the other stable state of the circuit. The condition for regeneration is a voltage gain from base to base greater than unity. A second trigger pulse will switch the circuit back to its original stable state.

Diodes D_1 and D_2 ensure that the negative trigger pulse is applied only to the transistor that is cut-off. Assuming again that Q_1 is cut-off, its base is positive while the base of Q_2, which is bottomed will be negative. Thus D_1 will be conducting, since it is forward biased, but D_2 will not, due to a reverse bias.

Thus a negative trigger pulse is transferred through D_1 to the base of Q_1 but does not reach the base of Q_2. When the circuit is in the other stable state the action of the diodes is also reversed.

The bistable circuit carries out one complete cycle for every two input pulses and during each cycle one pulse appears at each collector. The pulse repetition frequency of the input pulses may therefore be divided by any power of two by cascading the requisite number of binary counters.

Q.9.3

Write a short essay, illustrated by diagrams and formulae where relevant, on the following topic, Transistor switching and gating circuits.

[I.E.E., Nov. 1962, part (a) only]

A.9.3

The use of transistors in a bistable switching circuit is described in the previous question A.9.2.

One form of transistor gating circuit is shown in Fig. 9.3.1. This uses one resistor for each input and it is known as a transistor–resistor logic circuit.

If both inputs are at OV the transistor will be cut-off and the output is at nearly $-V_{CC}$. If a negative voltage is applied to

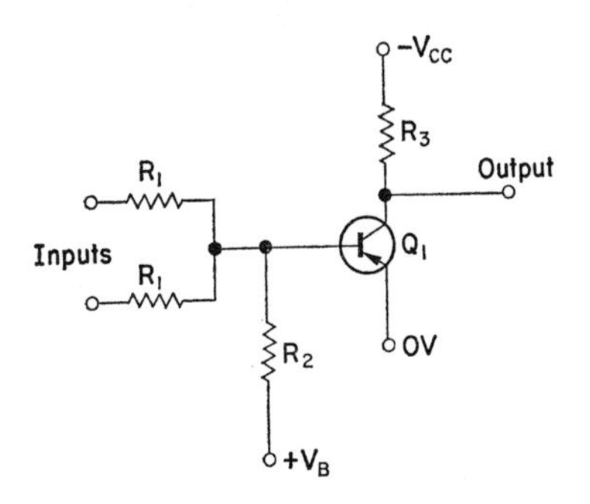

Fig. 9.3.1. Transistor–resistor logic circuit.

either or both the inputs, the transistor bottoms and the output rises to near zero. Since the output is the inversion of the input and a signal at either of the inputs results in a change in the output state the circuit is known as a NOT OR or NOR circuit. By suitable connection of NOR gates logic AND and OR gates can be realized.

This type of circuit is slow in operation. If only one input is present the transistor is turned ON with a base current:

$$I_{min} = \frac{V_{in}}{R_1} - \frac{V_B}{R_2}.$$

This results in a slow rise time due to the charging of the input capacitance. In the other extreme all inputs, possibly up to 10, may be present and this causes a large base current to flow. There will therefore be a long turn-off delay when all the inputs change to zero potential.

Desirable qualities for the transistors used in gating circuits are low bottoming voltage, short charge carrier storage time and a narrow spread of current gain.

Q.9.4

Draw the circuit of a binary counting stage (Eccles–Jordan circuit), employing two *pnp* transistors. Outline the mode of operation of this circuit.

By means of a block schematic explain how four such stages may be joined in tandem to give a decade counter.

[H.N.C. 3, 1963]

A.9.4

The first part of this question is covered in A.9.2.

Figure 9.4.1 shows four binary stages connected in tandem with feedback from the last stage to the second and third stages. One output pulse is obtained from P_4 for every ten input pulses applied to P_1.

Assume that initially the counter is reset so that there is no output from any of the four stages. After the seventh input pulse there will be outputs from the first, second and third stages. In a pure binary counter the eigth input pulse would switch the first three stages OFF and the last stage ON, producing an output from P_4. Due to the feedback, however, this output from P_4 switches P_2 and P_3 on again. The ninth input pulse will switch P_1 ON so that there are now outputs

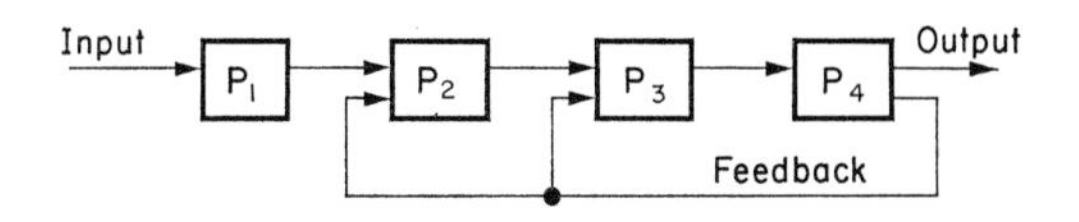

FIG. 9.4.1. Four binary stages to give decimal counting.

from all four stages. The tenth input pulse switches all stages OFF and the counter is returned to its initial state. Thus one output pulse is produced from P_4 for every ten input pulses, i.e. the counter is functioning as a decade counter.

Q.9.5

Explain briefly the physical operation and hence electrical characteristics of the tunnel (or Esaki) diode. Indicate how it may be used in a switching circuit and state the advantages and disadvantages of the device.

[C. and G. 5, 1961]

A.9.5

The physical operation may be explained using the energy level diagrams discussed in A.1.1. The energy level diagram for the heavily doped tunnel diode, with zero bias, is shown in Fig. 9.5.1. Due to the heavy doping the impurity levels form a fully energy band just below the conduction band for the *n*-type region, and an empty energy band just above the valence band for the p-type region. The depletion layer itself is very narrow, being of the order of 10^{-6} cm.

The application of a small forward bias (p positive with respect to n) reduces the energy barrier, and full impurity donor levels are separated from empty acceptor levels by a very narrow depletion region.

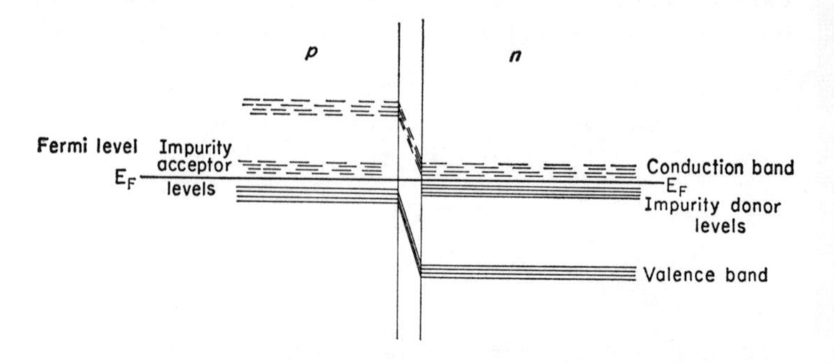

FIG. 9.5.1. Energy level diagram for tunnel diode with zero bias.

Under these conditions, electrons "tunnel" through the depletion region giving rise to a current much greater than the diffusion current. The "tunnel" current will increase with forward voltage until the maximum number of full donor levels are opposite the maximum number of empty acceptor levels. This occurs at about 50 mV.

Further increase in forward voltage reduces the "tunnel" current and at a few hundred millivolts the "tunnel" current is negligible compared with the normal diffusion current. The resultant characteristic is shown in Fig. 9.5.2.

The peak and valley voltages are determined by the material (gallium arsenide diodes have a larger negative resistance range than germanium) while the degree of doping and the junction area determines the current values. The valley current is usually about 10% of the peak current.

When used in switching circuits, as may be seen from Fig. 9.5.3, there are two stable conditions. On the characteristic a load line corresponding to a resistor R_S in series with the diode has been superimposed and the stable conditions *A* and *C* are separated by the negative resistance range. An input pulse

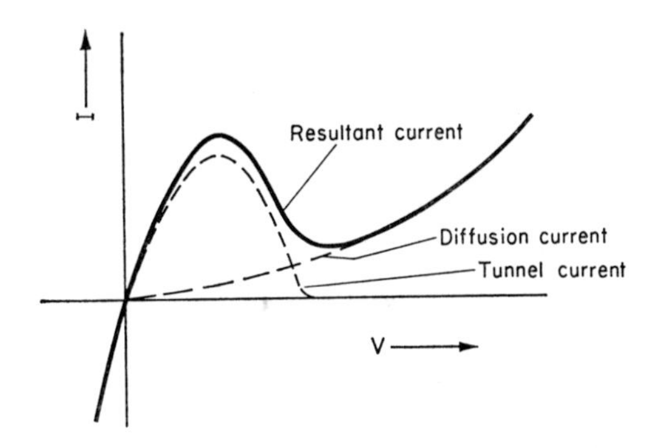

FIG. 9.5.2. Current–voltage characteristic of a tunnel diode.

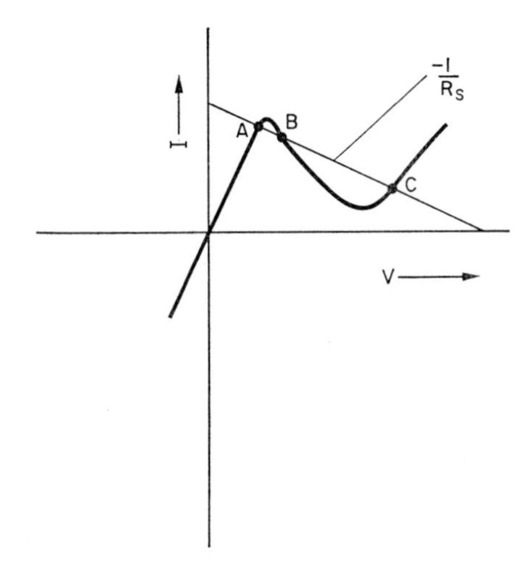

FIG. 9.5.3. Characteristic and load line for a tunnel diode, showing two stable states.

applied when in either of these states will carry the operation into the negative resistance part of the characteristic (e.g. B) and since this is unstable the device will switch to the alternate state.

The advantages of the device are:

(a) Physically small, giving low capacity to earth.

(b) Unaffected by temperature and radiation, since the hole–electrons pairs so produced are small in number compared with the effects of the heavy doping.

(c) Fast operation at high frequency with wide bandwidth and low noise.

The disadvantages include:

(a) The diode is not unidirectional due to the high reverse current and is not easily cascaded.

(b) The bias conditions must be closely controlled and the signal voltages that may be used are limited.

Q.9.6

Discuss the operation of transistors under saturation conditions in switching circuits. Describe a method of obtaining the advantages of saturation in an Eccles–Jordan type trigger circuit, whilst at the same time avoiding the disadvantage normally associated with saturation.

Figure 9.6.1 shows a basic trigger circuit. Estimate the maximum value of R_1 (in terms of R_2, R_3 and R_4) which may be used if saturation is to be avoided, no transistor parameters being given. State what assumptions it is necessary to make in the calculation.

[C. and G. 5, 1961]

A.9.6

In most transistors saturation occurs when the collector–emitter voltage v_{ce} is lower than the base–emitter voltage v_{be} and the collector base region is forward biased. The collector injects a charge into the base region and the collector current (I_C) cannot decrease until this stored charge has been removed.

Since a discrete change of state cannot take place in zero time, it follows that saturation should be avoided if a rapid change of state between ON and OFF is required.

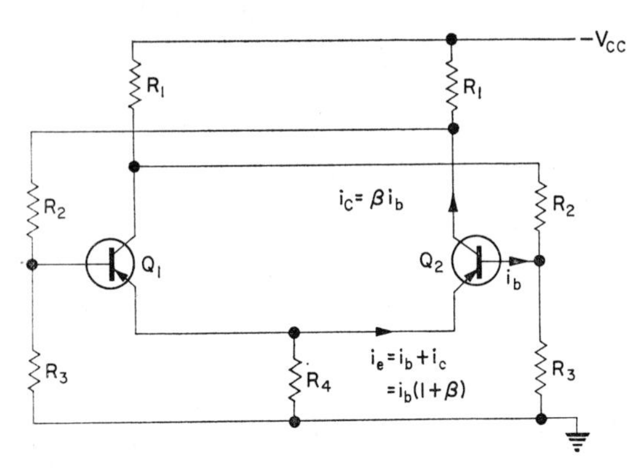

FIG. 9.6.1. Basic trigger circuit of Q.9.6.

A number of circuits have been developed to enable the advantages of saturation, namely, a definite ON characteristic to be obtained. These use additional diodes. The following derivation shows that the circuit in Fig. 9.6.1 may be made to have the same advantages.

If saturation is to be avoided v_{ce} must exceed v_{be}. The transistors are considered as being "perfect" current amplifiers and leakage currents are ignored.

Consider Q_2 in the ON condition:

$$v_{ce} > v_{be},$$

$$v_e > v_b,$$

i.e.
$$(\beta+1)i_b R_4 > \frac{R_3}{R_1+R_2+R_3} \cdot V_{CC}.$$

This assumes i_b is very much less than the standing current I through the resistance chain.

Also $$V_{CC} = \beta i_b R_1 + (\beta+1) i_b R_4 \quad \text{very nearly}$$
$$= (\beta+1) i_b (R_1+R_4) \quad \text{if} \quad \beta \gg 1.$$

Therefore $$(\beta+1) i_b R_4 > \frac{R_3}{R_1+R_2+R_3}(\beta+1) i_b (R_1+R_4),$$

i.e. $$R_4 > \frac{R_3(R_1+R_4)}{R_1+R_2+R_3}$$

or $$R_1R_4 + (R_2+R_3)R_4 > R_3(R_1+R_4),$$

i.e. $$R_2R_4 > R_3R_1 - R_1R_4$$

or $$R_1 < \frac{R_2R_4}{R_3-R_4}.$$

This gives the maximum value of R_1 if saturation is to be avoided.

Q.9.7

Give the circuit diagram, and explain the action, on an Eccles–Jordan bistable circuit.

Describe how such a circuit could be arranged as a binary counter stage operated by a train of negative input pulses. What limits the recurrence frequency at which such stage would operate?

[B.Sc. 3, 1960]

A.9.7

The first two parts of this question are covered in A.9.2. The frequency of operation is affected by:

(a) The input capacitance of the transistor. This is compensated for by capacitors C_1 and C_2 in the circuit of Fig. 9.2.1. For a high frequency of operation these should be chosen to give equal on and off switching times.

(b) Hole storage effects. This arises due to the time taken for holes to cross the base from emitter to collector. The transit time can be reduced by reducing the base width.

(c) Cut-off frequency of the transistor. As the frequency of operation is increased the gain of the transistor falls. If at some frequency the gain from base to base falls below unity, switching will not take place.

CHAPTER 10

Photo-electric Applications

Q.10.1. Silicon solar cell.
Q.10.2. Photo-transistor and photo-diode.

Q.10.1

Draw a cross-section sketch to show the construction of a typical silicon solar cell giving the significant dimensions.

Also sketch the current–voltage characteristics for such a cell when illuminated and in darkness and hence construct a simple equivalent circuit for a solar cell operating into a resistive load.

[H.N.C. 3, 1964]

A.10.1

The cross-section of a typical silicon solar cell is sketched in Fig. 10.1.1 with the significant dimensions.

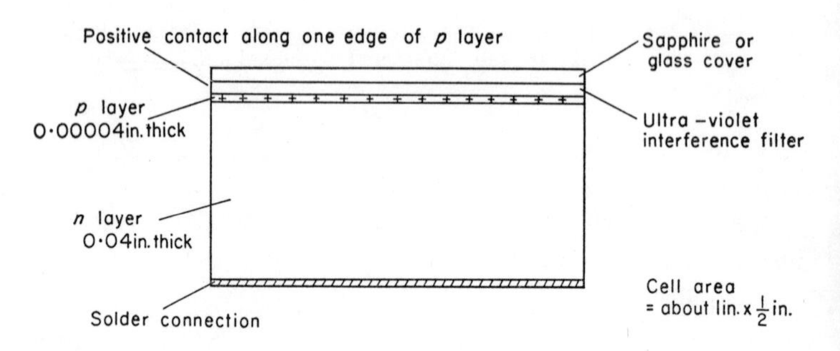

FIG. 10.1.1. Cross-section sketch of a solar cell.

The current–voltage characteristics for such a cell are given in Fig. 10.1.2, where graph (a) is for the cell in darkness and graph (b) is when the cell is illuminated, with a load line added to the graph. It can be seen from graph (b) that the solar cell,

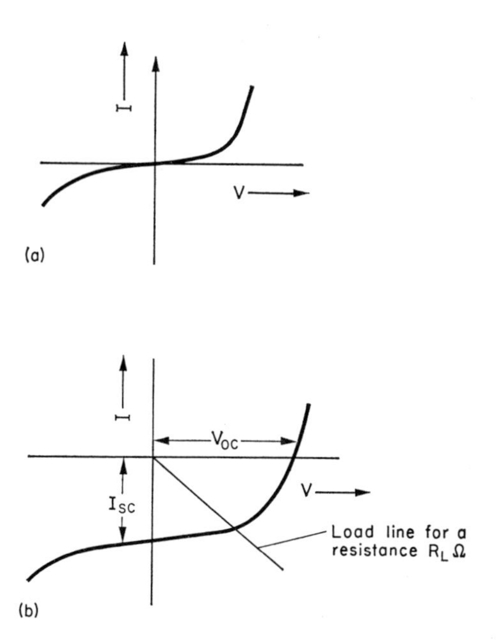

FIG. 10.1.2. Current–voltage characteristics for solar cell; (a) cell in darkness; (b) cell illuminated, with load line.

when illuminated, acts as a diode, together with either a current generator I_{SC} or a voltage generator V_{OC}. The shunt current generator is usually the more convenient form of equivalent circuit and will be chosen in this case. Also, when a resistive load is applied across the solar cell the available output voltage falls below the open circuit value. This is accounted for in the equivalent circuit by the internal series resistance R_{SE}. The complete equivalent circuit for an illuminated solar cell with resistive load is thus as given in Fig. 10.1.3.

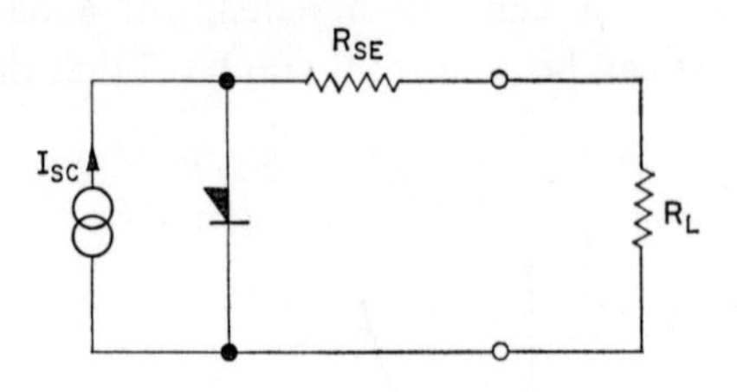

FIG. 10.1.3. Complete equivalent circuit for solar cell.

Q.10.2

Write a short account to summarize the differences in properties between a photo-emissive cell and a photo-transistor.

The photo-diode D_1 shown in the circuit of Fig. 10.2.1 has a constant sensitivity of 30 mA/lumen and negligible dark current. The gas-filled triode V_1 has a control ratio of 20 and, with zero grid bias, V_1 strikes at an anode voltage of 20 V. When V_1 conducts the anode current operates relay A.

Estimate the minimum illumination needed at the diode junction to operate the relay, ignoring any pre-striking grid current of the thyratron.

[H.N.D. 3, 1962]

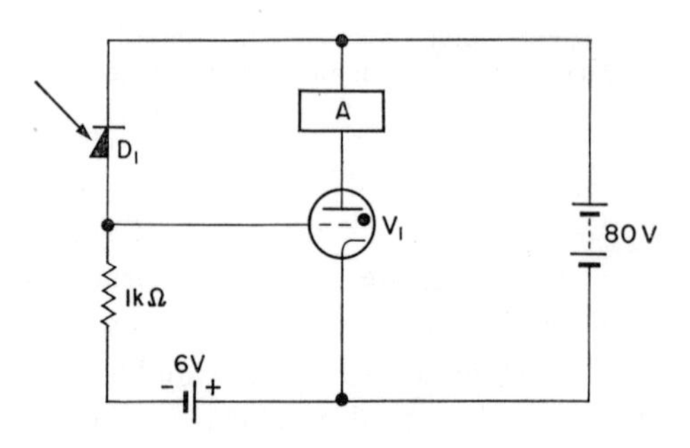

FIG. 10.2.1. Circuit with photo-diode and gas triode.

A.10.2

A photo-emissive cell is essentially a high impedance device, capable of supplying a small current into a load of several megohm. An anode supply voltage of the order of 100 V is required. The dark current from a photo-emissive cell is negligible and not temperature dependent. The ratio of output current to intensity of illumination is reasonably linear, such that the photo-emissive cell may be used to measure changes of light intensity.

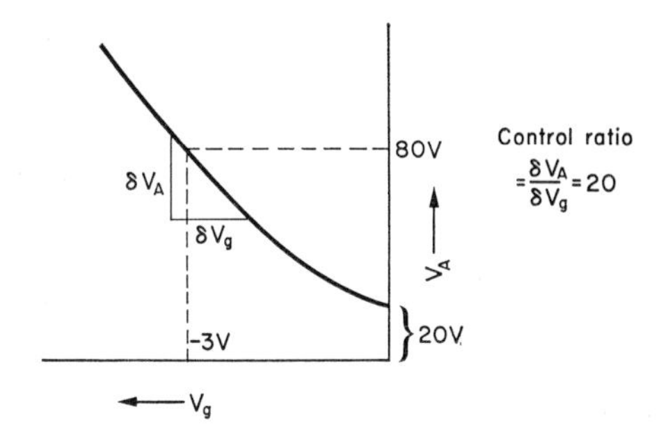

FIG. 10.2.2. Control characteristics of gas triode.

The small junction photo-transistor operates from a supply of about 10 V to supply currents of up to 100 mA, and can thus be used to operate a relay, or similar device, without any intermediate amplifier. The dark current from a germanium photo-transistor is often significant and is very temperature dependent. The characteristics are not sufficiently linear to give an accurate relationship between light intensity and output current. Thus the photo-transistor is best employed in ON–OFF applications.

If the gas triode in Fig. 10.2.1 has a control ratio of 20 as shown in Fig. 10.2.2 and conducts at 20 V on the anode with zero grid bias, then with a supply voltage of 80 V the triode will strike at -3 V on the grid. The effective grid voltage will be 3 V when there is a potential drop of 3 V across the grid resistor. This corresponds to a current of 3 mA through 1 kΩ. Hence, the minimum illumination needed at the diode junction to operate relay A is

$$\frac{3 \text{ mA}}{30 \text{ mA/lumen}} = \underline{0{\cdot}1 \text{ lumen.}}$$

CHAPTER 11

Special Applications

Q.11.1. Modulation and frequency changing with a diode.
Q.11.2. Microwave silicon crystal diode.

Q.11.1

A point-contact diode gives an output current of $i = \alpha v + \beta v^2$ when biased to a certain portion of the forward characteristic, where i is in mA and v in V. Show algebraically how this diode may be used for (a) amplitude modulation, and (b) frequency changing of an amplitude modulated signal.

Draw a circuit, for such a diode, suitable for the detection of a signal with a carrier frequency of 200 kc/s.

State appropriate values for the resistors and capacitors used.

[H.N.D. 3, 1963]

A.11.1

We will assume an input signal of the form $v = A \sin \omega_1 t + v_2$ where $A \sin \omega_1 t$ represents the carrier or local oscillator voltage and v_2 is the modulating or received voltage. By means of a correct choice of the form of v_2 it is possible to demonstrate either amplitude modulation or frequency changing.

In either case the output current from the diode is given by the expression

$$i = \alpha (A \sin \omega_1 t + v_2) + \beta (A \sin \omega_1 t + v_2)^2$$

or

$$i = \alpha A \sin \omega_1 t + \alpha v_2 + \beta A^2 \sin^2 \omega_1 t + \beta v_2^2 + 2\beta A v_2 \sin \omega_1 t.$$

(a) If v_2 represents a modulating signal voltage then $v_2 = B \sin \omega_2 t$. Then selecting terms from the expression for current, we have that,

$$v_{out} = K(\alpha A \sin \omega_1 t + 2\beta A v_2 \sin \omega_1 t), \quad \text{where} \quad K = \text{constant},$$

$$= K\alpha A\left(1+\frac{2\beta v_2}{\alpha}\right)\sin \omega_1 t,$$

i.e.

$$v_{out} = K\alpha A\left(1+\frac{2\beta B}{\alpha}\sin \omega_2 t\right)\sin \omega_1 t,$$

This expression represents an amplitude modulated wave, where the modulation index is given by $m = 2\beta B/\alpha$.

(b) If v_2 is taken to represent an amplitude modulated wave (usually the received signal) then

$$v_2 = C(1+m \sin \omega_3 t) \sin \omega_4 t.$$

Now, selecting the last term from the complete expression for diode current,

$$\begin{aligned} i &= 2\beta A v_2 \sin \omega_1 t \\ &= 2A\beta C \sin \omega_1 t(1+m \sin \omega_3 t) \sin \omega_4 t \\ &= A\beta C(1+m \sin \omega_3 t)\,[\cos(\omega_1-\omega_4)t \\ &\qquad -\cos(\omega_1+\omega_4)t]. \end{aligned}$$

Then

$$v_{out} = K_1 A\beta C(1+m \sin \omega_3 t)\cos(\omega_1-\omega_4)t$$

$$\text{where } K_1 = \text{constant}.$$

This assumes that the current is selected by a tuned circuit resonant at $f = (\omega_1-\omega_4)/2\pi$ c/s.

This last expression shows clearly that frequency changing has taken place, where the difference frequency term will usually be selected. That is, the output from the diode as a frequency changer will be of the form

$$K_1 A\beta C(1+m \sin \omega_3 t)\cos(\omega_1-\omega_4)\,t.$$

A diode detector circuit suitable for a carrier frequency of 200 kc/s is given with typical component values in Fig. 11.1.1.

Resistor R_1 forms the load for the diode, with the small capacitor C_1 acting as a high-frequency bypass. Resistor R_2 and capacitor C_2 act as a low cost high frequency filter, whilst capacitor C_3 blocks direct current from the output.

This solution has assumed a basic knowledge of modulation theory. This is available in many books on radio engineering (e.g. Terman, F. E., *Electronic and Radio Engineering*).

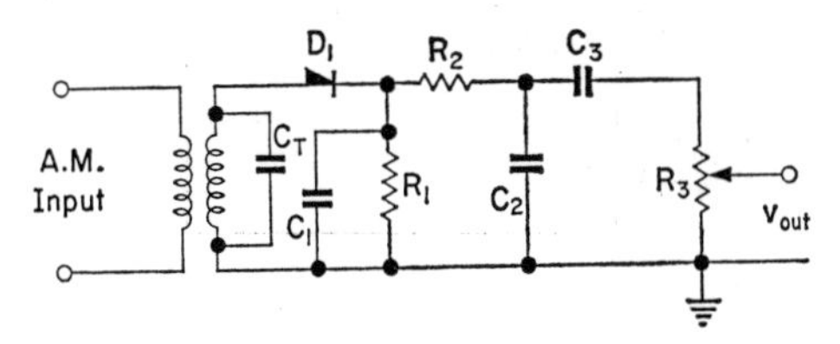

FIG. 11.1.1. Diode detector circuit for 200 kc/s operation.

Q.11.2

Sketch the circuit, in block schematic form, of a microwave frequency changer, employing a silicon crystal diode. Draw the voltage–current characteristic of such a point contact diode and hence explain the action of a diode mixer in UHF radio receiver.

What is meant by conversion loss and noise temperature ratio of a crystal mixer?

A.11.2

The circuit of a microwave frequency changer is given in outline as Fig. 11.2.1. The current–voltage characteristic of a crystal diode, which would be used in that type of circuit, is given as Fig. 11.2.2.

This characteristic shows that the diode may be considered approximately as a square law detector (in the first quadrant) it therefore follows, from A.11.1, that the diode will act as a frequency changer, or mixer, when fed with voltages from a local oscillator and the receiver aerial.

The gain G of a crystal mixer is the ratio of IF output power to RF input power. This is less than unity, and is usually expressed as a conversion loss, i.e. $L = 1/G$, where, at 3000 Mc/s, L is typically in the range from 4 to 8.

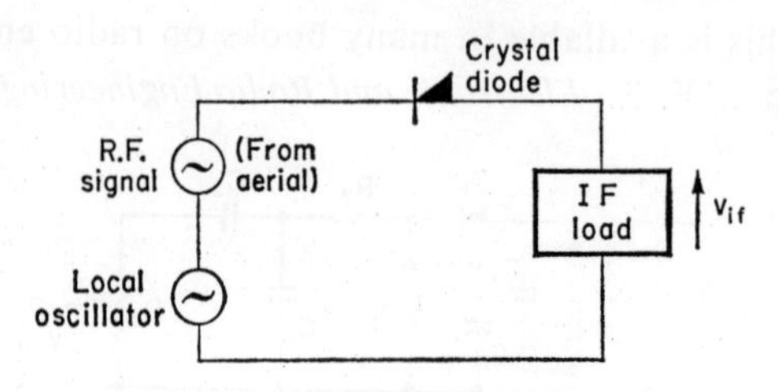

FIG. 11.2.1. Outline circuit of microwave frequency changer with a crystal diode.

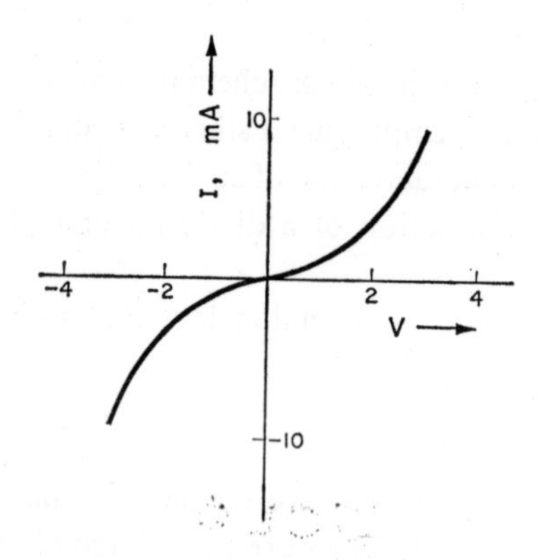

FIG. 11.2.2. Current/voltage characteristic of a crystal diode.

The noise temperature ratio t of a crystal is the ratio of the noise power generated in the crystal to that noise power which would be produced if the crystal was a pure resistance at room temperature. Usually this ratio t is in the range 1 to 3, and is often taken as $t=2$.

MADE IN GREAT BRITAIN